Agricultural Extension
Worldwide Innovations

Agricultural Extension
Worldwide Innovations

Ramesh Umrani
C.K. Jain

Oxford Book Company
Jaipur ▪ New Delhi

Published By

Oxford Book Company

Regd. Off. "Bony Residency", Gate No. 2
Opp. Tilak Public School, Vishwesv   Nagar,
Gopalpura Road, Jaipur - 302018
Ph: 0141 - 2761280, Telefax: 276
email - oxfordbookjpr@gmail.com
website - oxfordbookcompany.com

Delhi Office: 102, 1st Floor, Satyam House,
4327/3, Ansari Road, Darya Ganj,
New Delhi - 110002, India
Ph: 011-45652440

Published in India, 2025

ISBN : 9789380179162

Price : ₹ 2495/-

Typeset by Shivangi Computers, Jaipur

Printed and bound in India by Thomson Press Ltd.,
New Delhi

Contents

Preface

Agricultural extension was once known as the application of scientific research and new knowledge to agricultural practices through farmer education. The field of extension now encompasses a wider range of communication and learning activities organised for rural people by professionals from different disciplines, including agriculture, agricultural marketing, health, and business studies. Extension practitioners can be found throughout the world, usually working for government agencies. They are represented by several professional organisations, networks and extension journals. Agricultural extension agencies in developing countries have received large amounts of support from international development organisations such as the World Bank and the Food and Agriculture Organisation (FAO) of the United Nations.

India has the second largest extension system in the world in terms of professional and technical staff. According to recent estimates, India has more than 100,000 technical personnel in its extension system. India is in the process of transforming its agricultural extension and technology transfer systems to become more demand-driven and responsive to farmer needs. Agricultural extension continues to be in transition worldwide. Governments and international agencies are advancing structural, financial and managerial reforms to improve extension. Decentralisation, pluralism, cost sharing, cost recovery, participation of stakeholders in development initiatives and the decisions and resources that affect them—these are some of the elements in extension's current transition.

The present publication on agricultural extension is a thorough, practical handbook that explains the purpose, importance, and far-reaching effects of agricultural extension throughout the world. The information presented tells how to establish, communicate, and implement effective agricultural extension programs that address specific needs and long-term goals in any location. It is highly recommended for undergraduate and postgraduate students of agricultural or rural development studies in department of agricultural economics, geography, development studies, and rural sociology, agricultural extension educators and researchers, and other extension in government and associated organisations.

Ramesh Umrani
C.K. Jain

1

Introduction to Agricultural Extension

The term extension was first used to describe adult education programmes in England in the second half of the 19th century; these programmes helped to expand—or extend—the work of universities beyond the campus and into the neighbouring community. The term was later adopted in the United States of America, while in Britain it was replaced with "advisory service" in the 20th century.

The central task of extension is to help rural families help themselves by applying science, whether physical or social, to the daily routines of farming, homemaking, and family and community living. Extension involves the conscious use of communication of information to help people form sound opinions and make good decisions. The essence of agricultural extension is to facilitate interplay and nurture synergies within a total information system involving agricultural research, agricultural education and a vast complex of information-providing businesses.

Agricultural extension was once known as the application of scientific research and new knowledge to agricultural practices through farmer education. The field of extension now encompasses a wider range of communication and learning activities organised for rural people by professionals from different disciplines, including agriculture, agricultural marketing, health, and business studies.Extension practitioners can be found throughout the world, usually working for government agencies. They are represented by several professional organizations (such as APEN), networks (such as AGREN) and extension journals (such as Journal of Extension).

Agricultural extension agencies in developing countries have received large amounts of support from international development organisations such as the World Bank and the Food and Agriculture Organisation (FAO) of the United Nations.

In the US, an extension agent is a university employee who develops and delivers educational programs to assist people in economic and community development, leadership, family issues, agriculture and environment. Another program area extension agents provide is 4-H and Youth. Many extension agents work for cooperative extension service programs at land-grant universities. They are sometimes referred to as county agents or educators.

ORIGINS OF AGRICULTURAL EXTENSION

Men and women have been growing crops and raising livestock for approximately 10,000 years. Throughout this period, farmers have continually adapted their technologies, assessed the results, and shared what they have learned with other members of the community. Most of this communication has taken the form of verbal explanations and practical demonstrations, but some information took a more durable form as soon as systems of writing were developed. Details of agricultural practices have been found in records from ancient Egypt, Mesopotamia and China going back more than 3,000 years.

It is not known where or when the first extension activities took place. It is known, however, that Chinese officials were creating agricultural policies, documenting practical knowledge, and disseminating advice to farmers at least 2,000 years ago. For example, in approximately 800 BC, the minister responsible for agriculture under one of the Zhou dynasty emperors organized the teaching of crop rotation and drainage to farmers. The minister also leased equipment to farmers, built grain stores and supplied free food during times of famine.

The birth of the modern extension service has been attributed to events that took place in Ireland in the middle of the 19th century. Between 1845–51 the Irish potato crop was destroyed by fungal diseases and a severe famine occurred. The British Government arranged for "practical instructors" to travel to rural areas and teach small farmer how to cultivate alternative crops. This scheme attracted the attention of government officials in Germany, who organized their own system of traveling instructors. By the end of the 19th century, the idea had spread to Denmark, Netherlands, Italy, and France.

The term "university extension" was first used by the Universities of Cambridge and Oxford in 1867 to describe teaching activities that extended the work of the institution beyond the campus. Most of these early activities were not, however, related to agriculture. It was not until the beginning of the 20th century, when colleges in the United States started conducting demonstrations at agricultural shows and giving lectures to farmer's clubs, that the term "extension service" was applied to the type of work that we now recognize by that name.

COMMUNICATION WITHIN EXTENSION SYSTEMS

The term "extension" has been used to cover widely differing communication systems. Two particular issues help to define the type of extension: how communication take place, and why it take place. Early books on extension often describe a model of communication that involved the transmission of messages from "senders" to "receivers". As part of this model, senders are usually people in authority, such as government planners, researchers, and extension staff, while receivers are usually farmers who are relatively poor and uneducated. Although this model might include something called "feedback", it is clear that the senders are in control of the communication process.

The transmission model of communication is closely related to the idea that extension workers are the link (i.e. message carriers) between researchers (senders) and farmers (receivers). Extension programmes based on this model has been described as "paternalistic"; in other words, the actors in the communication process have a parent/child or teacher/student relationship. Other authors have used the term "top-down" to describe these programmes.

In many countries, paternalistic extension is gradually being replaced by more participatory approaches, in which the knowledge and opinions of farmers is considered to be just as important as that of researchers or government officials. Participatory approaches involve information-sharing and joint decision-making. The terms "interactive" and "bottom-up" have been used to describe these approaches.

The development of participatory extension requires a re-examination of the communication process. At the present time, no single description has replaced the transmission model that is referred to above, but two ideas are becoming widely accepted:

Communication in the context of participatory extension cannot usefully be described in a linear manner with distinct groups of senders and receivers. Instead, extension activities take place within a knowledge system consisting of many actors who play different roles at different times.

Although some actors in the knowledge system have more authority than others, communication usually involves a negotiation rather than a transmission. What takes place is a dialogue, with actors collaborating in the construction of shared meanings rather than simply exchanging information.

Although extension programmes have many different goals, most programmes fall into one of two basic categories:

- systems of communication that aim to change the behaviour of rural people
- systems of communication that aim to change the knowledge of rural people

There is, of course, a close relationship between knowledge and behaviour; changes in the former often lead to a change in the latter.

If government policy-makers, project managers or researchers direct the topics addressed and projects undertaken, then the purpose of extension is to change behaviour. This approach to extension has been variously described as directive extension, social marketing and propaganda.

If farmers and other rural people direct the extension towards their own needs, then the purpose of extension is changing knowledge. This knowledge helps rural people make their own decisions regarding farming practices. This approach to extension is closely related to non-formal education and conscientization.

PARADIGMS OF AGRICULTURAL EXTENSION

Any particular extension system can be described both in terms of both how communication takes place and why it takes place. It is not the case that paternalistic systems are always persuasive, nor is it the case that participatory projects are necessarily educational. Instead there are four possible combinations, each of which represents a different extension paradigm, as follows:

- *Technology Transfer* (persuasive+paternalistic). This paradigm was prevalent in colonial times, and reappeared in the 1970's and 1980's when the Training and Visit system was established across Asia. Technology transfer involves a top-down approach that delivers specific recommendations to farmers about the practices they should adopt.
- *Advisory work* (persuasive+participatory). This paradigm can be seen today where government organisations or private consulting companies respond to farmers enquiries with technical prescriptions. It also takes the form of projects managed by donor agencies and NGOs that use participatory approaches to promote pre-determined packages of technology.
- *Human Resource Development* (educational + paternalistic). This paradigm dominated the earliest days of extension in Europe and North America, when universities gave training to rural people who were too poor to attend full-time courses. It continues today in the outreach activities of colleges around the world. Top-down teaching methods are employed, but students are expected to make their own decisions about how to use the knowledge they acquire.
- *Facilitation for empowerment* (educational + participatory). This paradigm involves methods such as experiential learning and farmer-to-farmer exchanges. Knowledge is gained through interactive processes and the participants are encouraged to make their own decisions. The best know examples in Asia are projects that use Farmer Field Schools (FFS) or participatory technology development (PTD).

It must be noted that there is some disagreement about whether or not the concept and name of extension really encompasses all four paradigms. Some experts believe that the term should be restricted to persuasive approaches, while others believe it should only be used for educational activities. Paulo Freire has argued that the terms 'extension' and 'participation' are contradictory. There are philosophical reasons behind these disagreements. From a practical point of view, however, communication processes that conform to each of these four paradigms are currently being organized under the name of extension in one part of the world or another. Pragmatically, if not ideologically, all of these activities are agricultural extension.

PUBLIC SECTOR AGRICULTURAL EXTENSION

Diversified Strategies

Agricultural extension continues to be in transition worldwide. Governments and international agencies are advancing structural, financial and managerial reforms to improve extension. Decentralisation, pluralism, cost sharing, cost recovery, participation of stakeholders in development initiatives and the decisions and resources that affect them—these are some of the elements in extension's current transition.

Pressures to change

Public sector extension was severely attacked in the 1980s for not being relevant, for insufficient impact, for not being adequately effective, for not being efficient and, sometimes, for not pursuing programmes that foster equity. A critical turning point occurred that affected the way information transfer, heretofore considered the purview of public sector agricultural extension systems, was conceived and practiced. Not only did public extension systems come under public scrutiny and political attack but, as well, were confronted by heightened competitive interests from the private sector. Input suppliers and output buyers became increasingly active in instructing farmers in the processes and standards desired by particular markets.

Often enough, these information providers created demonstration plots and field trials, similar to public sector extension techniques but with a view to vertical technology transfer. In some cases, "contract farmers" turned into workers for the contracting companies. Despite the high returns associated with extension found in a number of studies there was a general feeling that public sector extension was overextended.

The scarcities of financial resources for extension and in some cases the lack of skilled manpower and dearth of organisational capacity led to major changes in ideological, economic and technical perspectives of agricultural extension. Also, the forces for worldwide structural adjustment resulting from massive debts by nations North and

South, the onslaught of conservative ideology emphasizing efficiencies over welfare, the accelerating reaction against subsidies in agriculture, all these contributed to the critical assessment of extension.

Thus, governments began to discredit and withdraw their commitment to extension. In their eagerness to evade extension's institutional problems and to relieve the larger economic problems, in a sense many governments chose to ignore extension.

Extension as a function

Extension is a nonformal educational function that applies to any institution that disseminates information and advice with the intention of promoting knowledge, attitudes, skills and aspirations, although the term "extension" tends to be associated with agriculture and rural development.

Its purposes may differ, from technology transfer by companies organised around specific, usually monocropping farm systems to problem-solving educational approaches to participatory programmes aimed at alleviating poverty and advancing community involvement in the process of development. Internationally, extension's institutional (and at present generally pluralistic) systems tend to differ from country to country. Most ministries of agriculture have an extension unit that deals mainly with crops and mixed agricultural systems, as well as separate technical divisions (livestock, forestry, fisheries, etc.) some of which also provide extension services.

During the 1970s and 80s, efforts were made to unify ministerial agricultural extension operations but with limited success. This same diversity and separation of agricultural extension activities exists in international organisations. Extension is multidisciplinary. It combines educational methodologies, communication and group techniques in promoting agricultural and rural development.

It includes technology transfer, facilitation, and advisery services as well as information services and adult education. It is dependent for success on other agricultural development processes such as marketing and credit services, not to mention economic policy and physical infrastructure. In short, it is a function that is dependent for success on other factors, including other services and institutions. In many cases its success depends on the ability to shift programme direction and development to stakeholders and programme users.

When systematically and effectively provided, extension is known to enhance social and economic development. Technological change and the knowledge system that underpins it, is a critical factor in development. Despite the difficulty of isolating its impact on agricultural productivity and growth from that of other factors, many studies have demonstrated the high economic returns of investments in agricultural dissemination. Investment in agricultural research and extension is thus a crucial input of agricultural growth. However, "agricultural extension services in developing countries are currently grossly under-funded to undertake the activities required for achieving food security while protecting the productive resource base in order to keep up with population and economic growth".

Reform strategies

Various and distinct public sector agricultural extension reforms have occurred since the mid-1980s.

The second (lower-right quadrant) represents change toward partial privatisation-whether through subsidies to producers who then are obligated to seek agricultural extension services from private providers, through grants to producer organisations or entities at local levels, or through public sector contracting with private or "third-sector" organisations to provide services.

The third (upper-right quadrant) refers to systems where the producer pays directly to the provider for services rendered. The fourth (lower-right quadrant) applies to situations where governments have either commercialised their public sector agricultural extension services, or have otherwise transferred these services in total to the private sector. The central rectangle refers to situations where governments employ several strategies (e.g., devolution, partial privatisation, and/or cost-recovery measures to foster extension activities for agricultural development). This phenomenon should not be confused with "institutional pluralism" that refers to the fact that various sectors are engaged in extension provision.

New linkages to other agencies and organisations in a multisectoral network of extension/communication providers will be necessary in a new vision of extension and rural development. Briefly, the upper-left quadrant indicates a partial or total shift of authority from the central authority to lower levels of authority or subgovernment. Deconcentration is an example of shifting authority to branch offices or regional institutes. Devolution is when the funding and delivery of extension services is shifted to the state level (Brasil, Mexico) or to the municipal/district level (Bolivia, Colombia, and The Philippines).

In the lower-left quadrant, three strategies are cited. In the first, the central authority selects to provide subsidies to small farmers on condition of their contracting with private sector providers for extension services. In the second, the central authority contracts directly with private entities to provide services to producers. An interesting example is the case of Honduras where the government has contracted with a coordinating service outside its own country to then contract with private extension providers within Honduras.

These two strategies-subsidising farmers to contract with the private sector and contracting with the private sector-are strategies aimed in part at enabling the private sector to provide extension services. The third strategy in the lower-left quadrant is relatively recent; it involves providing grants to communities directly for them to undertake development projects. This World Bank initiative seeks to motivate communities to take ownership for their own development. Still, the community projects are not necessarily for developing extension services.

In the upper-right quadrant is the strategy of fee-based extension service. An example of cost-recovery for extension services is Ecuador's Programme for Modernisation of Agricultural Services (PROMSA); however, this programme involves both the public and the private sector in the financing and delivery of services and therefore incorporates reform measures associated with the lower-left quadrant.

Again, this is an example of a mixed system and, indeed, Ecuador's approach to extension involves subsidising as well as cost-recovery programmes. This quadrant underscores the fact of agricultural information's commodification. The lower-right quadrant refers to countries that have totally commercialised or privatised their public sector agricultural extension services. This quadrant also includes those countries that selected to withdraw completely from provision of extension services and support for extension education, and generally shifted authority to national and international NGOs. This quadrant reflects the general withdrawal of some countries during the late 1980s and throughout the 1990s from public sector involvement in goods and services and the consequent shift of authority to the private sector or third-sector organisations.

Consensus on lessons learned

At the international workshop on "Extension and Rural Development" jointly hosted in November 2002 by the World Bank, the USAID and the Neuchatel Group, a number of policy recommendations were put forward. They are central to the issues of extension, rural development and food security. Some recommendations for Poverty reduction are listed below:

1. Poverty reduction should be the focus of public funding whether provided by public sector staff or contracted out. Given the emphasis on poverty reduction and the increasing knowledge intensity of rural income-generating activities operating in a globalising economy, the role of public funding is likely to increase. Government should focus public financing on the poor.
2. The emerging view of extension is not that of a service or system, but of a knowledge and information support function for rural people. Rural knowledge and information needs are diverse; thus there are benefits from having a range of

providers to deliver advice, technology innovations, and facilitation services. Government should view extension within a wider rural development agenda.

3. Extension policy should be designed with an inventory of the actors and an assessment of the quality of the services rendered before deciding on any reform. It is important to have a strategy for a national extension system, but this requires a country-led vision and political support independent of donor agendas although in line with country-driven processes, e.g. each country's Rural Poverty Reduction Strategy. Government should define extension policy for a pluralistic system.

4. New approaches will take years to be fully institutionalised. Long-term resolution must be taken within a widely shared vision at different levels-international, national, regional and community. Government should make long-term commitments.

5. Some type of coordinating body for the various participants in extension needs to be considered to provide a common framework in which all actors can operate. Policies and mechanisms need to harmonise behaviour and strategies (e.g., minimum levels for co-financing, prioritisation and area selection). Government should develop a stakeholder coordinating mechanism.

6. Capacity building at all levels is critical. Capacity building and institutional strengthening require funding to widen the pool of qualified service providers and ensure strong links with and modernisation of the various components of the formal and non-formal agricultural education system. Government should build capacity of rural organisations (RPOs), the public sector, and service providers.

7. The private sector will play an increasingly important role in rural knowledge systems, but total privatisation is not feasible, even for commercial agriculture. The appropriate mix of public and private roles can only be determined through piloting and learning from experience. Government should be realistic about the limits of fully private extension.

8. Extension services should be a part of a larger decentralisation agenda that engages local government units and grassroots organisations. Once the decision is taken to revisit the potential of extension/communication as a major thrust toward a food security initiative, then the next step for government is to develop a nationwide strategy for extension and rural development.

The above sentences offers a valuable set of recommendations for governments to consider with respect to agricultural extension and rural development.

Following two decades of criticism and public sector reform of extension services—sometimes involving severe downsizing or even withdrawal from the provision of services—a new recognition of the role of the public sector in promoting extension is resulting in the advancement of diverse strategies of agricultural extension. This adoption or build-up of diverse strategies for delivery of extension services points toward the tendency of the future.

Extension for Poor Rural Families

Poverty worldwide is excessive. In many countries as much as two-thirds of the population are "dirt poor", with minimum access to basic needs including adequate nutrition, clean water, proper sewage, and health care.

The rural sector deserves immediate and considerable attention. At least half the world's poor live in rural areas, and the majority of these rural poor work the land. More than 60% of the poor in Latin America and the Caribbean live in rural areas, and their poverty is more extreme than that of the poor in urban areas; the magnitude and heterogeneity of their situations are such that it is difficult to determine what precisely are the causes and to therefore assign possible solutions. Likewise, the factors affecting food security vary widely.

Some governments have engaged in processes of devolution with the aim of ensuring that local governments possess the resources and authority to confront poverty. This appears to be an important step toward involving a broader array of participants in the fight to reduce poverty and food insecurity. At the same time more carefully differentiated extension strategies are required if governments are to reduce poverty among the rural poor because poverty is a multidimensional phenomenon. To date, various approaches recognise diverse livelihoods, e.g., the "sustainable livelihoods approach" (LSA) and the "farming systems approach" (FSA).

These are in addition to the Special Programme for Food Security, which fosters a production/irrigation approach while incorporating elements of both LSA and FSA. These different programmes tend to overlap in their goals. However, LSA places emphasis on vulnerability and tends to be a social approach.

FSA focuses on the farm household and is a more technical approach. Special Programme for Food Securityconcentrates on food security and income generation related to agriculture, and is of particular interest to the present document because it is essentially an agricultural extension programme that focuses on the rural poor.

Enhancing food security and income generation

Among the various FAO and other international programmes aimed at assisting developing countries to reduce food insecurity, the SPFS began by supporting food security in Low-income Food Deficit Countries (LIFDCs), but as of 2003 the programme was operating projects in over 70 developing countries. In general, the programme sets out to help developing countries to:

- control water resources through micro schemes that protect from the vagaries of the weather (drought and flooding) that cause serious fluctuations in annual output levels;
- boost the crop, livestock and aquaculture productivity, including diversification and intensification of production by small farmers so that they can feed their families and secure a surplus to increase earnings;
- identify and find measures for responding to the socio-economic constraints on the production, marketing and processing of agricultural commodities;

SPFS projects differ according to the particular situation of the geographic region, its natural resources and the characteristics of the people involved in the project. It is also affected by other elements such as project leadership and government policies.

The programme lays out five major corporate strategies:

(1) Contributing to the eradication of food insecurity and rural poverty;

(2) Promoting, developing and reinforcing policy and regulatory frameworks for food, agriculture, fisheries and forestry;

(3) Creating sustainable increases in the supply and availability of food and other products from the crop, livestock, fisheries and forestry sectors;

(4) Supporting the conservation, improvement and sustainable use of natural resources for food and agriculture; and

(5) Improving decision making through the provision of information and assessments and fostering knowledge management for food and agriculture.

SPFS projects centre in particular on the diversification of income opportunities for the rural poor, and the responsiveness of policy interventions to the requirements of disadvantaged groups.

The main task is to enhance small farmer livelihoods by improving their agricultural output and thereby contributing to their food security and income generation. At the same time, the projects promote related activities such as participatory forestry, gender approaches and, not least, encourage the creation of agriculture-related facilities such as the building of greenhouses and processing centres, and development of commercial activities among the poor.

The projects utilise participatory approaches and methods to support food security and income generation. On-site examination of two SPFS projects in Guatemala confirms that project interventions are being carried out successfully and that the ancillary attributions regarding local institution building are clearly evident in project implementation. While SPFS may not have taken hold everywhere nor been successful in every instance, the point is that it has the capability of significantly improving the lives of the people in the places where its leadership and resources are adequate.

Programme development and future expectations

SPFS implementation takes place in two phases. Phase I is a micro-economic phase consisting of four major components: (1) water control through small systems of water collection, irrigation and drainage, (2) intensification of crop production systems, (3) diversification of production systems into small animal production, "artisanal" fisheries and aquaculture, and (4) analysis and resolution of socio-economic constraints to food security. Phase II, i.e. the macro-economic phase, has three dimensions:(a) an agricultural sector policy reform to overcome socio-economic constraints, (b) an agricultural investment programme to address infrastructure, and (c) the preparation of feasibility studies of bankable projects designed to ensure bilateral and multilateral financing.

Core features of the SPFS strategy are national ownership with the participation of farmers and other stakeholders at all stages of the programme's conception and implementation, priority given to small farmers, environmental awareness, integrated and multidisciplinary approach, emphasis on modernisation of low-cost simple technologies, and social equity (gender and role of vulnerable groups).

In all SPFS projects, national agricultural extension services are actively involved in introducing improved technologies, forming men and women farmer groups, and organising field demonstrations. However, their contribution remains limited, and varies according to the extent of their involvement in planning and decision-making, and their strengths and weaknesses. Even though the purpose of the SPFS is not strengthening of a country's extension system, yet it is imperative that extension services be strong enough for meaningful implementation, monitoring, impact assessment and sustainability of the SPFS projects. This is especially true in case of those SPFS countries where extension services have become weak as a result of decentralisation or structural adjustment programmes.

SPFS leadership in the FAO hopes that SPFS projects will eventually become national programmes in the countries where they are successful. In those countries where agricultural extension programmes no longer exist, have been badly reduced, or have become weak, SPFS implementation staff could become the equivalent of a national extension service, in this case specifically serving poor populations who work the land. In contrast, where government has successfully reformed agricultural extension system, utilising a diversified strategy for extension development, the SPFS will form an important component within the overall extension framework.

RURAL DEVELOPMENT

Investing in agricultural development is good business. Despite the increase of larger farms and the disappearance of small farms, small farming situations still provide the basis for much of contract farming in the developing world. Investing in rural development is also good business. While agricultural development and rural development very much interact with each other, they nevertheless need to be distinguished, especially with respect to the extension function and also with regard to discussions of poverty and poverty alleviation.

Agriculture is, as de Janvry and Sadoulet make clear, only one of four basic potential paths out of poverty in Latin America. Nonetheless, their analysis is likely to be the reality in other regions of the world. The pathways they cite include:

Agricultural Path

This is the case of landholders who have sufficient natural capital endowments, and for whom market, institutional and policy contexts allow for the profitable use of these assets. OECD's *West Africa Long Term Perspective Study* foresees major changes to the agricultural sector over the next twenty years. Predictions are that a small number of larger, commercial farm operators will emerge, able to invest in new technology, sell into world markets and compete with imports from elsewhere. In drier, lower potential areas, patterns of farming and levels of productivity will change less markedly, with increasing diversification into a range of off-farm sources of income, including migration. This agricultural path has been the traditional focus of integrated rural development interventions.

The result has been mixed success in reducing poverty and has generally produced unsustainable programmes. However, new programmes such as the SPFS and other extension services focusing on food security and income generation hold promise for catalysing rural development through advancement of agricultural development among the poor.

Multiple-activity Path

This path is dominant among rural households in Latin America. Yet it has been generally unrecognised and unsupported, except for local interventions with limited success at scaling up. Major rethinking about the institutional design of rural development is needed to incorporate the off-farm income dimension into these strategies. It should also be clear that the agricultural part of this path is not the same as that of the agricultural path itself. For farmers involved in multiple activities, agriculture is often a part-time endeavor.

The household's off-farm activities are often undertaken to generate liquidity for farm expenditures. The off-farm part of the multiple-activity path is not the same as that for fully landless households that have more flexibility in the labour market and often better location relative to the sources of effective demand, as compared to rural residents who are part-time farmers. A typical observation, thus, is that part-time farmers achieve on average levels of household income that are lower than those of landless workers.

Assistance Path

Well supported in the urban sector, this path is profoundly neglected in rural areas. It applies to the structural poor caught in poverty traps who need permanent income transfers to reach the poverty line, and to households in transitory poverty who need access to safety nets to avoid decapitalisation of productive assets and irreversible adjustments to shocks.

Exit Path

Although sometimes ignored in agriculture, migration from rural areas—as an exit strategy—has been the dominant factor in reducing rural poverty in Latin America. Remissions by migrants to Latin America alone are estimated to amount to two billion dollars per annum.

These paths are not necessarily mutually exclusive. However, households that come out of poverty can usually be identified with one or the other. Identifying which path offers the greatest promise is important for designing differentiated rural development interventions that can best help poor households escape poverty.

The rural sector, however, remains under-served, despite the major and positive changes in the 1990s in the way governments and development agencies approached rural development and poverty.

The gap between the rich and the poor appears to have broadened over the last thirty years. In Latin America, according to Ferreira and Walton, slow economic growth and technological development that benefit only the highly educated are the biggest causes for the increase of the gap in the region. The region, which in the past had a competitive advantage because it offered cheap labour, now suffers as the industrialised world seeks highly qualified employees who know how to handle new technologies. As a result, millions in Latin America are unemployed.

Governments with the assistance of international organisations are beginning to promote various decentralised programmes, including subsidiarity-providing communities and rural producer organisations with the potential to develop their own programmes for local development. An interesting approach to local development is the

World Bank's promotion of projects that empower rural people via community-driven development (CDD), encouraging communities toward self-determination. These CDD projects assist communities in the formulation of proposals that are then reviewed and if accepted, then funded.

In other cases, governments have begun to initiate efforts toward nationally integrated and multisectoral extension networks to combat food insecurity. These incipient national system networks include public, private and third sector organisations as well as international projects aimed at food security goals. Governments have forged partnerships with other sectors of society, including multisectoral providers of extension and information services, to foster conditions to end hunger. These governments expect a food security strategy to increase domestic food security and also eventually to facilitate inter- and intra-regional trade in food items.

Agricultural Extension and Rural Extension

Just as the distinction needs to be made between agriculture as an aspect of rural development and other, non-agricultural elements as aspects of rural development, so a distinction needs to be made between "agricultural" extension and "rural" (or rural development) extension.

Distinguishing between agricultural extension and rural extension

The distinction is needed. Extension agents trained in agronomy and livestock development are unprepared to take on the various roles arbitrarily assigned to them by those eager to solve development problems on paper. Agricultural extension agents have already been commandeered to take on tasks involving construction of postharvest on-farm infrastructure, marketing and processing, farm management and the organisation of farmers into special agricultural interest groups. Preparation in these areas requires in-depth knowledge, positive attitudes and special skills training, and selectivity as to which agents are likely to respond well to such training.

Tasks associated with "rural extension" include micro-enterprise development, nonformal literacy education, family planning, nutrition, health and other rural, non-agricultural areas needing attention. In would be easy to state simply that these tasks must be assigned to either a separate or integrated extension staff. Certainly, it cannot be assumed that specialists in agriculture will overnight become specialised in these other, equally demanding, practices.

A multi-sectoral extension strategy and network of providers covering both agricultural and non-agricultural extension, involving different agencies and organisations, is necessary. A separate or integrated extension staff, one part of which is dedicated to agricultural activities and the other involved in non-agricultural activities, could be established in the near future. One answer-as a first step-is development of a communication strategy that operates to serve both of these two major populations in rural areas: the people who work the land for a living and those who do not.

Communication for rural development

Strategies that include communication for rural development as a significant aspect of agricultural and rural development are sorely needed. Efforts in this direction are being made, but governments have yet to recognise fully the potential of this factor in promoting public awareness and information on agricultural innovations, as well as on the planning and development of small business, not to mention employment opportunities and basic news about health, education and other factors of concern to rural populations, particularly those seeking to improve their livelihoods and thereby enhance the quality of their lives.

Rural development is often discussed together with agricultural development and agricultural extension. In fact "agricultural extension" is often termed "rural extension" in the literature. In contrast, rural development includes but nonetheless expands beyond the confines of agriculture, and furthermore requires and also involves developments other than agriculture.

Accordingly, government should consider the establishment of a communication policy that while supporting agricultural extension for rural development also assumes the role of a "rural extension" service aimed as well at diffusing non-agricultural information and advice to people in rural areas. A communication policy would aim to systematically promote rural communication activities, especially interactive radio but also other successful media such as tape recorder and video instructional programmes.

Computers and the Internet may not yet be accessible to rural communities but they serve the communication intermediaries and agricultural extension agents who provide information to rural populations. Other devices such as cell phones hold considerable promise for the transfer and exchange of practical information. For reaching the final agricultural and basic needs information users in rural areas today, radio is the most powerful and cost-effective medium. However, other traditional and modern communication methods are equally valuable, depending on the situation and availability, like face-to-face exchanges (via demonstration and village meetings); one-way print media (such as, newspapers, newsletters, magazines, journals, posters); one-way telecommunication media (including non-interactive radio, television, satellite, computer, cassette, video and loud-speakers mounted on cars); and two-way media: (telephone, including teleconferencing, and interactive (Internet) computer).

Information and communication technologies (ICTs) have proved to be important for Internet users and for the intermediate users who work with the poor. Pilot experiences show that various media are valuable for assisting agricultural producers with information and advice as to agricultural innovations, market prices, pest infestations and weather alerts.

ICTs also serve non-farming rural people with information and advice regarding business opportunities relating to food processing, wholesale outlets and other income-generating opportunities. In the case of non-agricultural rural development interests, a communication for rural development policy would aim to promote diffusion of information about non-agricultural micro-enterprise development, small business planning, nutrition, health and generally serve to provide useful, other-than-agriculture information.

By its very nature as mass media, communication for rural development can provide information useful to all segments of rural populations. However, it would serve as a first effort toward advancement of "rural extension" services and activities aimed at rural development concerns beyond those of agriculture. Thus, extension and communication activities would be expected to work in tandem, allied in the common cause of supporting income-generating activities, both agricultural and non-agricultural.

Communication as related to extension services immediately suggests several avenues of mutual support. For example: these would include national services relating to extension and communication, specialised extension communication services, extension services promoted by producers, commercialised extension services, and mass media extension-related services.

A similar orientation toward other aspects of rural development information and technical advice is evident considering the de Janvry-Sadoulet rural development pathways and other related rural development needs such as information and assistance with health problems, most notably Human Immunodeficiency Virus/Acquired Immunodeficiency Syndrome (HIV/AIDS) in case of sub-Saharan Africa.

Rural extension and radio need to be more purposely connected. Radio, according to contemporary specialists, is under utilised at present. While ICTs and their connection to radio hold promise for the future, some consider radio to be "the one to watch". In this connection, regional networks are being launched. Examples are The World Association for Community Radio Broadcasters (AMARC) and the Latin American Association for Radio Education (ALER). Global initiatives have begun: Developing Countries Farm Radio Network (DCFRN) and UNESCO Community Media Centres.

FOOD SECURITY

Food security is often defined in terms of food availability, food access and food utilisation. Food availability is achieved when sufficient quantities of food are consistently available to all individuals within a country. Such food can be supplied through household production, other domestic outputs, commercial imports or food assistance. Food access is ensured when households and all individuals within them have adequate resources to obtain appropriate food for a nutritional diet.

Access depends upon income available to the household, on the distribution of income within the household and on the price of food. Food utilisation is the proper biological use of food, requiring a diet providing sufficient energy and essential nutrients, potable water, and adequate sanitation. Effective food utilisation depends in large measure on knowledge within the household of food storage and processing techniques, basic principles of nutrition and proper childcare. "Global agriculture currently produces ample calories and nutrients to provide all the world's people healthy and productive lives", according to the USAID document. "However, food is not distributed equally to regions, countries, households and individuals.... Improved access to food-thr ough increased agricultural productivity and incomes-is essential to meet the food needs of the world's growing population. "

Successful food security and poverty-oriented programmes not only assist poor rural populations to produce more and diversified products but to produce a surplus that can be marketed and thereby generate income for the purposes of improving quality of life through improved diet and nutrition, investment in productive activity, and as collateral for credit to purchase inputs and/or other supplies to enhance agricultural or non-agricultural enterprise.

Government's first task is to provide "the public goods needed by societies to remain peaceful and prosperous, goods that are unlikely to be produced in sufficient quantity by private markets alone or by non-governmental institutions". Public goods are goods with benefits that are available to all and not diminished in their availability even when consumed. Paarlberg argues that provision of public goods is only the first task of government, and would include some non-public goods as well; in this category he includes food security.

In the area of food security, one such good might be a supply of cheap food made available to the poor through a public food distribution system. In other cases the pursuit of food security might even require that private goods be taken from a traditionally privileged category of citizens, with or without compensation, for redistribution to disadvantaged citizens. In still other cases food security might require government action to reduce racial prejudice or gender inequity.

When governments signed the WFS pledge in 1996 to cut food insecurity in half, they acknowledged the importance of food security to the general welfare and as a public benefit. Still, all sectors must be involved in this fight against hunger, as the FAO Director-General has stressed. It is important that nearly 500 national and international private sector representatives attended the WFS. However, the role of the public sector is still central to this fight, as private sector investment in the developing countries is strongly influenced by the governance, legal framework, financial services, and transport, communications and energy infrastructure. Food security is, and will likely remain for some time into the future, the outstanding "public good" obligation of countries worldwide.

Investment in extension

A review of rural investment is taking place among international organisations indicating increasing concern for broad-based and other-than-agriculture entrepreneurial development in rural areas. The Inter-American Development Bank (IDB) is reviewing its priority areas of investment to include farmers without agricultural potential and the landless poor, as well as farmers with agricultural potential, and has launched a major effort in favour of the development of the rural economy and poverty reduction. The World Bank and the Neuchatel Group also highlight the challenge of poverty and its alleviation, as well as the Millennium Development Goals.

When systematically and effectively provided, extension is known to enhance social and economic development. Technological change, and the knowledge system that underpins it, is a critical factor in development, according to the World Bank. In spite of the difficulty of isolating its impact on agricultural productivity and growth from that of these other factors, studies have demonstrated the high economic returns of investments in agricultural research and dissemination, with returns typically above 40 percent.

Investment in agricultural research and extension is thus a crucial input of agricultural growth. At present, however, agricultural extension services in developing countries are grossly underfunded to undertake the activities required for achieving food security while protecting the productive resource base in order to keep up with population and economic growth. Successful SPFS projects such as in Pakistan, demonstrate that food security and income generation can go together. Wellmanaged projects emphasize diversification of products as well as income opportunities for the rural poor, and the responsiveness of policy interventions to the requirements of disadvantaged groups.

The project professionals begin by examining the agricultural situation, rural development constraints and problems of food security. Their work is both direct and "indirect" in that they work directly with farmers but also maintain indirect relations with a variety of local institutions, both for support and feedback to these institutions. Their activities may range from participatory forestry enhancement, animal diversification, and crop intensification-all with a view to promoting sustainable livelihoods. These SPFS extension professionals work with participants to better manage soil and water, to diversify their animal production, intensify their agriculture, and to organise themselves around their special agricultural interests, such as the building of greenhouses and processing facilities.

Investment in extension for women in agriculture

Women are the backbone of domestic work and the household economy as well as in most production, processing and storage activities, and marketing of agricultural products. This reality is often ignored, overlooked, or forgotten in putting together agricultural extension programmes. Evidence from observations in the field highlights the role of women in developing various activities of crucial importance to production and marketing of agricultural products.

Administrators in developing countries need to disaggregate data regarding the agricultural and agriculture related activities of women and men. Field personnel need to ensure that women are recognised for their contributions and that their potential is encouraged through inclusion in decision-making and that support is provided to them with respect to organising them for the purposes of production and other agricultural development activities.

REFERENCES

Adams, M. 1982. *Agricultural Extension in Developing Countries,* Longman.

Brunner, E. and Hsin Pao Yang, E. 1949. *Rural America and the Extension Service.* Columbia University

Maunder, A. 1973. *Agricultural Extension: A Reference Manual* (1st Edition), FAO

Saville, A.H. 1965. *Extension in Rural Communities: A Manual for Agricltural and Home Extension technician Workers.* Oxford University Press.

2

Development of Agricultural Extension

The use of the word "extension" derives from an educational development in England during the second half of the nineteenth century. Around 1850, discussions began in the two ancient universities of Oxford and Cambridge about how they could serve the educational needs, near to their homes, of the rapidly growing populations in the industrial, urban area. It was not until 1867 that a first practical attempt was made in what was designated "university extension," but the activity developed quickly to become a well-established movement before the end of the century. Initially, most of the lectures given were on literary and social topics, but by the 1890s agricultural subjects were being covered by peripatetic lecturers in rural areas.

The growth and success of this work in Britain influenced the initiation of similar activity elsewhere, especially in the United States. There, in many states, comparable out-of-college lectures were becoming established by the 1890s. During the first two decades of this century, the extramural work of the land-grant colleges, concerned with serving the needs of farm families, was to expand dramatically and become formally organised; but the use of the term "extension" continued and has persisted as the designation for the work. The society sought to do this largely through producing low-priced publications and establishing local committees throughout the country "for extending the object of the Society". During its twenty years' existence, agricultural topics were well covered in the society's publications. Similar, albeit short-lived, societies were also established before 1840 in several other European countries, India, China, Malaysia, and the United States.

CONDITIONS FOR THE GROWTH OF AGRICULTURAL EXTENSION

Apart from the importance of farmers and agriculture in the society and economy concerned, several conditions appear to be necessary for the initiation and organised development of agricultural extension work. The prime condition is that information has been assembled, systematised, and made available on good or progressive or new agricultural practices suited to a particular environment, and is based on either the accumulation of experience or findings from research. Second, this information is used, among other things, to educate professional agriculturists who may further enlarge or refine this body of knowledge or become active promoters and disseminators of it. Third, an appropriate administrative or organisational structure exists by and within which the dissemination activities may be established and conducted. Fourth, there is a legislative or some other official mandate or influential proponent which prescribes or enables that agricultural extension work is desirable and must occur. Fifth, there are invariably a variety of antecedents which have attempted protoforms of agricultural information and advice dissemination.

In addition, the incidence of critical situations, such as famine, crop failure, soil exhaustion, or altered economic conditions or relationships, may create an immediate cause for initiating the organisation of extension work. All or several of these conditions have been present in the evolution of modem forms of agricultural extension.

The need for agricultural and rural information and advisory services is likely to intensify in the foreseeable future. In much of the world, agriculture faces the challenge of keeping pace with rapidly increasing population with few reserves of potentially cultivable land. Farmers will have to become more efficient and specialised. From government perspectives, whatever priority is given to production, extension will remain a key policy tool for promoting ecologically and socially sustainable farming practices.

Some of the most promising recent developments in extension methodology have occurred where the key agenda is environmental or is concerned with equity, for example in the need for the joint management of forests by professionals and local forest users and in integrated pest management.

A consistent theme running through the innovative approaches being used, such as participatory rural appraisal, is a fundamental change in what are the respective roles of extension agent and clients. The agent is no longer seen as the expert who has all the useful information and technical solutions; the clients' own knowledge and ingenuity, individually and collectively, are recognised as a major resource; solutions to local problems are to be developed in partnership between agent and clients. Since the scale at which extension support is required is thus often larger than the individual farm, extension workers need new skills of negotiation, conflict resolution, and the nurturing of emerging community organisations.

The future is also likely to witness a reversal of recent trends towards bureaucratisation within hierarchical extension services and a reduction in their levels of public funding. Moreover, a rapid increase can be expected in the use of information technology in support of extension. The forces for change in these areas will come from four main directions.

Economic and Policy Climate

With the collapse during the past decade of socialist forms of economic organisation, the (dominant) role of the public sector in national economies has become questionable, with a strengthening trend to reduce levels of public spending. Thus government extension services and those which are largely publicly financed are, and will continue to be, under pressure to become more efficient, to reduce their expenditure and staff, and to pass on the costs of provision to their clients who directly benefit financially. This is particularly the case in countries where the farm population forms a small minority and agricultural production is in surplus.

The case is weaker, but not absent, in less developed countries where farming households form a high proportion of the total population and where increasing food production is still important. Thus charging clients for services is likely to become more widespread, while governments will find it attractive to contract out the operation of services to the private or the voluntary sector.

Systems Knowledge

A recognition of the locale-specific nature of farming systems and the agricultural information systems which support them is an important source of the pressure towards the debureaucratisation and devolution of extension services. This recognition also implies that extension workers and farmers be jointly involved in the verification and adaptation of new technology, and thus that the extension workers respect farmers as experimenters, developers, and adapters of technology and devote more energy on communication within their local areas. The devolution of extension services to become local organisations is a reasonable corollary of this. Developments in mass media technology, already apparent over a decade ago, will continue to support this localisation of extension effort.

Information Technology

The continuing rapid development of telecommunications and computer-based information technology (IT) is probably the biggest factor for change in extension, one which will facilitate and reinforce other changes. There are many possibilities for the potential applications of the technology in agricultural extension. IT will bring new information services to rural areas over which farmers, as users, will have much greater control than over current information channels.

Even if every farmer does not have a computer terminal, these could become readily available at local information resource centres, with computers carrying expert systems to help farmers to make decisions. However, it will not make extension workers redundant. Rather, they will be able to concentrate on tasks and services where human interaction is essential—in helping farmers individually and in small groups to diagnose problems, to interpret data, and to apply their meaning. The future will call for more able, more independent, more client-oriented extension workers.

The emphasis will be on the quality of interaction between agent and client rather than on the movement of "messages" through a hierarchical system. Flexibility and adaptability will be seen as virtues rather than aberrations. Paradoxically, these trends will bring us full circle to the early manifestations of modem extension in Europe. The itinerant agricultural teachers, unencumbered by large bureaucracies and tall hierarchies, will find their modem counterparts in the computer-carrying extension workers who are at ease helping farmers to identify the information they need in order to realise the potential of their farming operations. Looking back, we can regard the period from 1970 to 1995 as a necessary but expensive stage in the evolution of extension systems, after which extension agents were able to settle down to their main task—bringing together the expertise of farmers and the best available scientific knowledge to develop farms and local agricultural economies.

ALTERNATIVE APPROACHES TO ORGANISING EXTENSION

Extension, as the organised exchange of information and the purposive transfer of skills, is a rather recent phenomenon. Obviously, transfer of information and skills has existed since the emergence of permanent agriculture. Today's practice is different in that the process is dominated by organisations, and its scope has extended from disconnected local events to a complicated, large-scale, and even worldwide activity.

Extension approaches are presented in terms of their most important organisational forms and their respective goals. The goal system reflects the power positions of various groups of actors. Therefore, without an understanding of the historical development and of the interest groups involved, present achievements and shortcomings of extension approaches cannot be evaluated. It is assumed that different forms of organising extension are per se neither "good" nor "bad." Rather, extension services must be judged against their proper goals. The one universal yardstick, however, is their service function to the rural communities. Extension which is not in touch with and does not significantly contribute to improving the life situation of its clientele has lost its legitimisation.

Extension Goals

Goals lead the actions of individuals, groups, and organisations. While pointing towards a future state, they are influenced if not determined by past experiences. They reflect the interests of their stakeholders and differ, therefore, according to specific life situations, power positions, and development philosophies.

The prominent features of a system, such as its organisational structure, the choice of clientele, its operational design, and the methods used, are directly influenced by its set of goals and must be evaluated in terms of their contribution to goal achievement. Main actors within the extension system are the members of rural communities, extension and other development personnel, researchers, and staff of commercial or public service and support organisations. Empirical evidence shows a variety of forms in which interaction among these groups is institutionalised.

The variety of forms suggests a similar variety of goals, and either could be used to classify extension approaches. In practice, however, one finds an almost inseparable mixture of goals inhibiting a clear-cut classification. It seems more appropriate, then, to use a broader category, namely, selectivity with regard to clientele, and treat the respective goals as a continuum. The two end points of this continuum would be marked as technology transfer and human resource development, suggesting either a rather narrow technical or a broader socioeconomic view of development.

Technology Transfer

Until the end of the eighteenth century, farming techniques developed gradually and steadily over centuries with few qualitative leaps. Colonialism and imperial expansion introduced innovations—the spread of maize, tobacco, potatoes are striking examples—but experimentation and dissemination of knowledge were basically at the local farm level. The rise of agricultural sciences has induced dramatic changes in this respect. Increasingly, new technology has been created outside the actual farming sector by public sector research organisations. More recently, private firms in industrialised economies find agricultural technology research and development a highly profitable business.

For decades the research-extension-farmer linkage, especially in developing countries, was based on a rather simple model. In order to achieve development, "modem" research results had to be transferred to the "traditional" farmer, and extension seemed to be the appropriate means to do so. The general faith in science and the commitment to modernisation led to discrediting indigenous knowledge. Although this view is still held by many administrators, researchers, and extension agents, it is now being seriously questioned. Farming systems research and the "rediscovery" of farmers' knowledge have shown that "improved technology is a package of inputs and practices that usually comes from many sources".

The reexamination of the conventional view on agricultural knowledge cannot, however, result in questioning the important role of research as the source of new technology. For developing countries, one observes that the accelerated growth and spread of problems—such as the degradation of marginal land—surpass the problem-solving capacities of the local population. What is called for is a setting of new priorities and the building of knowledge systems based on problem solving rather than on information transfer.

Human Resource Development

The concept of human resource development is much broader than that of technology transfer, though both are closely interrelated. Increasing complexity not only of technology but also of the life situation of farmers even in remote areas demands new skills. With the help of these skills, rural women and men "acquire a better insight into the network of problems and recognise the alternative solutions available".

Traditionally, teaching the basic skills of literacy and numeracy has not been an extension activity. The limited success of literacy programmes in poor countries has drawn attention to nonformal education in which extension has an important part to play. Whereas in most cases this would require a coordinated effort of different organisations of which extension is but one, human resource development may also be regarded as a genuine extension content. Extension may substitute over a certain period activities such as vocational education that are not yet in place, but more important will be the teaching of managerial and organisational skills that will enable farmers to increasingly solve their own problems.

Human resource development thus aims at what may be called "critical competence." Extension clients know what to ask for, they can evaluate the appropriateness of technical information, they are responsible decision makers. Persons with this qualification exist in every rural community, and they will be the ones who actively seek further assistance. One important task of any extension system will therefore be to extend human resource development to underprivileged groups with less access to formal or vocational education—women farmers, rural youth, and generally small farmers in remote areas.

Alternative Ways of Organising Extension

The goals of extension may vary, as was shown, within the overall system as well as between different extension organisations. In addition, specific objectives may sometimes contradict each other. While smaller systems may come close to pursuing a consistent set of objectives or reconciling conflicting interests, large-scale organisations must work on a compromise basis. In this respect, Axinn's principal observation is of particular importance: "The success of an agricultural extension programme tends to be directly related to the extent to which its approach fits the programme goals for which it was established". The alternatives to organising extension demand choices on various levels:

- Public versus private
- Government versus nongovernment
- Top-down (bureaucratic) versus bottom-up (participatory)
- Profit versus nonprofit
- Free versus cost-recovery
- General versus sector
- Multipurpose versus single purpose
- Technology driven versus need oriented

In practice, extension organisations everywhere pursue the overall goals of technology transfer and human resource development, though the emphasis will differ. Within each organisation there is a mix of objectives, and within countries there is often a mix of organisational patterns. When presenting an overview on the most important patterns, we will be using a well-established terminology, though the grouping is different. We will differentiate between approaches that, at least in principle, target all persons in rural areas engaged in farming and those that purposely select clientele according to specific criteria.

Role of Extension Staff

Person-to-person communication has traditionally been the most important form of information transfer. Print media as well as radio and television were of a supplementary nature because they frequently lacked a target group or location specificity and information was not up-to-date. Revolutionary changes in communication technology have dramatically increased the speed and quality of information transfer and changed the role of extension workers in industrialised countries. Electronic communications systems may in part replace personal visits, and one of the major tasks of any agent will be to link her or his clients with other suppliers of information.

Notwithstanding the fact that their use is not restricted to industrialised countries, the fascination with modem communication means tends to obscure the fact that most extension personnel—90 per cent of which are located in developing countries—are working under extremely difficult and disadvantageous conditions. In fact, little has changed during the last two decades to remedy the basic ills, and the field agent is still the weakest link within the system.

There is a wide discrepancy between organisational goals and the potential of even well-trained and dedicated field staff. Early Indian community development activities covered close to fifty areas, all to be administered by the local village-level worker.

Fieldwork in most developing countries is characterised by conditions that foster low morale: lack of mobility, virtually no equipment, and extremely low salaries. For many extension workers, tapping additional income sources is a question of physical survival. Quality performance is further impeded by low educational qualifications and lack of advancement possibilities.

While working conditions of extension personnel have deteriorated, expectations with regard to their role are increasing. They are no longer to be simply transmitters of technical knowledge. They are to practise participatory methods, recognise and respect gender issues, identify indigenous needs and problem solutions, and serve as a link to the world outside the village, to name but a few of the present topics. The emerging role is closer to that of a "socio-economic community worker" than a technical expert, but their training is insufficient for either.

The situation sketched above is well known and documented. The sheer dimension of the problem surpasses, however, the capacities of poorer countries, notably in Africa. Foreign-funded projects have addressed the issue in a piecemeal fashion and have often drained nonproject areas of personnel. Staff reductions on a national level have not even secured the status quo. Neither approach has solved the basic dilemma: insufficiency of material and human resources to reach universally accepted societal goals. Having to count on their own resources for extension, many countries will not be in a position to implement technology transfer, much less the more demanding strategy of human resource development. Regardless of specific extension approaches, there is no alternative to a strong international commitment to strengthening and revitalising extension personnel resources.

AGRICULTURAL EXTENSION IN CHANGING RURAL CONTEXT

Agricultural extension, whether public or private, operates in a context or an environment that influences the organisation, form, and content of transfer activities. The dominant characteristic of that context is change. Because the changes affect all aspects of extension, the context should be examined and understood so that extension can be better managed.

The first places extension within a system of agricultural technology generation and transfer. Two major sections follow. One describes the macro-context factors: agroecological, infrastructural, policy, political-economic, and sociocultural. The other looks at the institutional context, namely, other institutional actors involved in activities related to agricultural extension such as research and transfer, education and training, input supply, and credit.

The usefulness of a systems approach for understanding and analysing agricultural technology generation and dissemination is widely acknowledged. Although they are sometimes criticised because they are so abstract, systems analyses offer holistic vantage points for understanding the factors that impede or enhance the two-way flow of technology and information between farmers and the public organisations that constitute the system.

The model used here identifies an agricultural knowledge system consisting of four components set in a larger context. The components are technology generation, technology transfer, technology utilisation, and agricultural policy. The organisations that constitute the components, as well as others in the system environment, influence each other in complex ways. Such publicly funded systems are established by governments to improve the conditions of life and well-being of rural and urban populations and to increase agricultural productivity.

The functions and linkages related to the flow and feedback of technology and information in the system define the components. In the public sector, agricultural research organisations and extension are major actors, although commercial companies and NGOs are commercial companies and NGOs are also involved. The way these are divided among organisations differs from country to country.

Technology generation consists of planning, administration, and implementation of research activities that develop, assess, adapt, and test improved agricultural technology for farmers and other users. In the public sector, these tasks, as well as some dissemination work, are carried out by agricultural research organisations.

Technology transfer further evaluates and adapts research outputs for users and then widely disseminates the knowledge and inputs to different target adopters—farmers of different categories, private companies, and so on. In many countries, government extension does both types of transfer, although the emphasis is on knowledge. Private organisations are also involved in both types of transfer, although farmer coverage is more limited and input transfer may play a bigger role.

The technology utilisation component encompasses the users of the agricultural technology, mainly farmers. User awareness, adaptation, and adoption of improved technology from various sources affect farm-level productivity and profitability and, ultimately, economic growth at the national level. Interaction and feedback between users and research and transfer organisations improve cooperation and the relevance of technology.

The policy component relates to government development goals and strategies, market and price policies, and the levels of resource investments in the system. Various government bodies play a role in setting development policy. Technology development and transfer organisations are affected by the policy in fundamental ways.

The division of generation and transfer tasks organisations and sectors differs from country to country, as does the nature of the policy and utilisation components. However, the flow of technology can be mapped as it moves between or is influenced by organisations constituting the agricultural technology system. Flow analysis shows the impediments to generation and transfer and the influences of other institutional actors and highlights contextual issues.

Macro-factors

Agroecological

Because the natural environment strongly influences extension planning and operations, extension should respond to the technological needs of farmers in different agroecological zones. The variation represented by agroecological zones in a given country can be substantial. Differences in temperatures, rainfall, soil types, evapotranspiration, and so forth are reflected in the diversity of farming conditions and production systems found.

Extension planners face some difficult choices because of the need to respond to the diverse technology and information needs of farmers from many different zones and, at the same time, to satisfy a requirement for extensive countrywide coverage of the rural population. For example, extension resource investments can be determined by farmer population concentrations, potential productivity of selected agroecological zones, or a combination of both.

Political-economic

The political and economic environment affects extension in many ways. One of the most significant factors is a country's stage of economic development. Another is the level of government investment in public sector extension. This is influenced by the presence or absence of a structural adjustment programme, the degree of economic dependence on agriculture, and the proportion of the population economically active in agriculture as opposed to industry. The percentage of resource-poor smallholder farmers influences the type of technology to be transferred, particularly if government is concerned with the equitable coverage of all categories of farmers.

Politics has another, infrequently identified impact on public extension organisations. Political shifts at the national level often result in changes in extension personnel, management, and programmes. High turnover of top managers undermines management experience and continuity in leadership.

Sociocultural

In many countries, sociocultural factors are leading constraints to the effectiveness of extension. Language differences and illiteracy can impede the communication of improved technology unless they are taken into account. The division of labour between the sexes can differ along cultural lines and influences the nature of farming systems in different regions. In many countries, the men are employed off-farm, leaving the farm operations to women. In extension organisations, under representation of women on the extension force means that the production responsibilities and needs of women at the farm level may not be adequately addressed.

Cultural differences among farmers, as well as differences in their resource endowments, also need to be taken into account. In particular, these are reflected in land-use strategies. Pastoral herders, for example, will require different types of subject-matter expertise, and extension will need to use different strategies to transfer technology to them than to permanent field agriculturalists. The resource endowments of different categories of farmers also affect technology adoption levels. Subsistence farmers adopt mainly low-cost technologies. For this reason, extension work that focusses on cultural practices and affordable technologies may be more appropriate in countries with large numbers of resource-poor farmers.

Policy

The policy component of an agricultural technology system can enable or limit extension in ways beyond the reach of extension managers. The principal areas of influence are price signals to farmers and decisions by government that affect public agricultural development organisations.

Agricultural Development. Policy-making bodies of government set development goals and objectives such as achieving food security or surplus agricultural production to stimulate economic growth, or providing health care and education for rural development. It is the task of government extension organisations to help meet these goals by formulating subordinate objectives and strategies to achieve them.

The amounts invested by government and the development community in agricultural development influence the pace and scale of effort. In many instances, spending ceilings for public institutions have been imposed by a structural adjustment programme agreed to by government. Extension planning needs to take these limits into account, especially when dealing with the question of staffing, which has major budget implications.

The endowment levels for public agricultural extension are set by government policy and planning bodies and impose limits on expenditures. This can hamper extension efforts in basic ways. The operating as compared to salary budget amounts are often inadequate, with negative effects. For example, a lack of money to buy vehicles and fuel undermines the mobility of extensionists. In addition, if salaries are too low, extension cannot attract or hold qualified staff, and services will suffer. Careful planning, guided by a strategy and programme of action, can improve the use of resources and the effectiveness of extension work within government funding limits.

Market Intervention. Governments set policies on consumer and producer commodity prices, subsidies for inputs, credit availability, import substitution, export earnings, food self-sufficiency, and natural resource management. These send direct and indirect price signals to farmers and influence their production decisions. High fertiliser prices, for example, cause farmers to use less of this input. Extension organisations should be aware of market signals to ensure that they are recommending technologies that meet farmers' current needs. They also require feedback from farmers to ensure the continued relevance of their activities.

Commodity prices influenced by government policy act as incentives or disincentives for farmer production. If there is no profit incentive for production of a specific crop, there is little point in developing or transferring improved technology related to that crop. For example, policies that favour the import of cereal grains at concessionary prices on the international market discourage incountry production of those crops. Technology generation and transfer outputs that focus on such crops are not likely to interest farmers. The examination of the policy context is important and allows organisations to avoid wasted effort and resources.

Infrastructure

Infrastructure, particularly the condition of transport, communication, and market facilities, affects both farmers and extension. The capacity to move people, inputs, and produce and to send and receive information influences extension activities and capacity. Market infrastructure itself can be lacking or inadequate.

Transportation. There may be many areas that cannot be reached by road, or transport vehicles may be in short supply. In either case, farmers under these conditions are difficult to reach with improved technology, and they will have problems transporting inputs and farm produce. For example, input supply outlets can be far apart, and excessive transport distances can make it difficult or impossible for traditional farmers to adopt improved technology.

Communications. Communication infrastructure can impose additional constraints for extension organisations. Farmer access to mass media such as publications, radios, or television may be limited, thus reducing the options available to extension for communicating its messages. At the same time, extension itself may have little or no access to telephone and radio services for long-range communications. This can severely hamper its ability to organise and carry out field operations.

Two aspects of a country's media organisations, both print and audio-visual, affect the flow of extension messages to farmers. One is the attitudes and subject interests of media managers responsible for programming for rural audiences. The other is the organisational climate, especially morale. It is crucial that extension consider the capacity of mass media organisations (newspapers, magazines, radio, and television) as part of its strategy and plans for communicating with farmers.

Institutional Factors

Public agricultural extension organisations in most countries have the task of providing a two-way flow of improved technology and information between research and users, primarily farmers. They operate in an institutional environment that includes other public and private organisations active in agriculture. In particular, those other actors involved in generating and transferring agricultural technology must be examined and understood to improve extension's effectiveness and efficiency.

Research

Agricultural research organisations are extension's closest institutional partners in technology generation and transfer. While these functions are also undertaken by private organisations, public sector organisations have the task of generating technology relevant for all categories of producers and agroecological zones. The way research is structured and organised, and the planning and management of research-extension linkages, can limit or enhance extension's effectiveness.

Agricultural research organisations in developing countries confront many problems. These include lack of financial resources, acute shortages of well-trained scientists, lack of farmer feedback to ensure relevance of research results, lack of access to external sources of knowledge, inadequate research facilities and equipment, low staff morale, and inadequate operating budgets, staff incentives, and remuneration. Few of these can be addressed by extension managers, but they can impede the generation of technology, resulting in fewer research outputs for extension to transfer. An understanding of research's problems is an important step in planning extension activities and coordinating them with research.

Poor linkages between research and extension are major constraints in technology flow in many countries. The linkage problems are of two basic types: those affecting feedback from farmers to research and extension and those relating to coordination and cooperation between research and extension. In both cases, extension managers can improve the situation by developing a linkage strategy and allocating responsibilities and adequate resources for linkage tasks.

Education and training

Educational organisations that train extensionists are important elements in the institutional context for extension. The work of universities and training institutes in particular has a significant impact on extension organisations. The content of their curricula as well as the numbers and qualifications of their graduates are limiting or enabling factors in any country. In many cases, communication between extension and education organisations is poor. As a result, extension commonly has staffing problems. It is not unusual for extension organisations to have posts that are either vacant or filled by underqualified personnel.

Inadequate numbers and qualifications of staff remain a difficult problem for public sector extension organisations. Salaries and benefits are rarely competitive with those of comparable private and public enterprises, resulting in low morale and high staff turnover. Education levels may be quite low, especially for farmer contact staff. The ability to attract and retain qualified extension staff is limited in most countries by civil service salary scales established by other agencies of government.

The situation can be improved by establishing staffing and training plans. The staffing plan inventories current human resources, identifies staffing gaps, and projects staffing needs over a specified time. The training plan identifies specific types of training (in-service and formal) required to fill skill gaps in human resources and to cover staffing needs for planned operations. The additional step of coordinating training needs with the educational organisations is needed.

Input supply

Farmers need inputs to increase production, but access to these is often poor in less developed countries. While inadequate transport and marketing infrastructure are often at the root of the problem, there are certain aspects that can be addressed by extension.

Genetic technology. Among the major outputs of the technology generation subsystem is improved genetic material. New plant and animal varieties with higher yields or resistance to pests or diseases become available. However, farmers need to know when the variety is released, how it performs under farm conditions, and where to obtain seed or breeding material. Extension is responsible for disseminating this information through appropriate mass media and contact methods. But a common constraint on the flow of technology is the availability of genetic material. In some countries, extension is also involved in the multiplication and distribution of seed. Establishing effective linkages with others involved in the process can also help ensure that genetic material is available for farmers.

Agrochemicals and other inputs. The performance of new varieties is often improved by, or even dependent on, the availability of agrochemical and other inputs at the farm level. Recommendations for fertiliser types and amounts suited to local soil conditions, for animal feed mixtures and practices, and for the control of plant and animal pests and diseases constitute an indispensable part of extension messages. Lack of access to this information and materials prevents yield maximisation, so extension organisations need to ensure that farmers are informed on availability and use. If access is limited because of external factors, extension should plan its campaigns accordingly, with attention given to low-input recommendations.

Private companies and nonprofit organisations also provide advice to farmers on agrochemical and other input use. Extension organisations need to develop communication and coordination linkages with these actors because excessive use of agrochemicals can harm human health and the environment. Conservative recommendations and alternative approaches, such as integrated pest management, are in the interests of the public and can be promoted by extension through its farmer contacts and other linkages.

Credit

Access to credit is one way to improve farmer access to new production technology and increase productivity. Farmers' ability to purchase inputs such as improved seed and fertiliser is particularly important. If appropriate technology is available but not being used by farmers, then the way credit is handled by government may be part of the problem.

Understanding the credit context—government and bank policies, availability of credit, and the institutional relationships involved in its delivery—is important for extension. At a minimum, the existing credit situation should be examined so that factors affecting the adoption and use of technology can be identified. These include inequitable access to credit, insufficient amounts of it, and overlap of transfer activities due to credit institution involvement in extension work. The knowledge is useful for extension managers in targeting farmers and in coordinating extension objectives with credit institutions.

Farmer organisations and other NGOs

There is growing involvement of the private sector, both nonprofit and commercial organisations, in agricultural research and extension. While such organisations may have limited research objectives and restricted regional coverage targets, they are partners in technology generation and transfer. Private corporations such as seed and agrochemical companies play a key role in developing some types of technology, providing inputs, and advising farmers on their use. Ideally, extension should reach agreements with these actors so that duplication of effort is minimised and conflicting messages to farmers are avoided. At a minimum, their activities should be noted and an attempt made by extension to catalogue and use their successful methods and outputs.

Farmer organisations, particularly grass-roots organisations, are part of the utilisation component. They offer an effective channel for extension contact with large numbers of farmers, as well as opportunities for participatory interaction with extension organisations. Feedback on farmer needs, production problems, and the results of adoption from such groups will be increasingly important considerations.

ECONOMIC CONTRIBUTIONS OF AGRICULTURAL EXTENSION

Agricultural extension programmes are quite diverse from an international perspective. Most are managed as public sector agencies, usually located in the ministry of agriculture, but some are located in other ministries such as education or rural development. Many are managed by nongovernmental organisations (NGOs). Many private firms and private organisations conduct extension programmes. Even within the most typical organisational structure, where extension is part of the government's ministry of agriculture, there is great variation in the degree of decentralisation of management of extension services. In some countries, extension is decentralised, as in India, where it is a state subject.

In most developing countries, however, governmental services are highly centralised, with varying forms of regional and subregional units designed to serve local areas. Further, there is great variation in the skill level and agricultural competence of field staff. In some systems, field staff have little formal technical training in the agricultural sciences. In some cases, this is dictated by a village worker philosophy, in others by local language demands. But, in most cases, it simply is the result of the decisions to expand agricultural extension programmes rapidly during the 1950s and 1960s, when few highly trained agriculturalists were available.

Finally, this diversity of skills, management systems, and objectives has changed over time in many countries. Perhaps the major changes in the management and design of agricultural extension systems over the past four decades is associated with the training and visit (T&V) system introduced in the 1970s by Benor, Harrison, and Baxter and implemented in many countries with World Bank lending support. Given this diversity, broad generalisations about the economic contribution of agricultural extension to agricultural development are not feasible. Many situation-specific factors impinge on the effectiveness of extension programmes.

The fact that substantial reform and redesign of many extension programmes has taken place indicates that some of them were perceived by their supporters to have been less than fully effective.

Conceptual Foundation for Extension Impact

Two conceptual themes are relevant to extension impact. The first is the awareness-knowledge-adoption-productivity (AKAP) sequence. The second is the "growth gap" interrelationship between extension, schooling, and research

The AKAP sequence

It is convenient to visualise extension as achieving its ultimate economic impact by providing information and educational or training services to induce the following sequence:

- A: Farmer awareness
- K: Farmer knowledge, through testing and experimenting
- A: Farmer adoption of technology or practices
- P: Changes in farmers' productivity

Changes in farmer behaviour will be reflected in quantities of goods produced, the quantities of inputs used, and in their prices. These, in turn, can be measured as "economic surplus," which is the added value of goods produced from a given set of inputs made possible by the extension activities. Studies of extension impacts have measured farmer awareness, knowledge, adoption, and productivity. Not all studies have examined all parts of the sequence. Most have shown a statistical relationship between the quantity of extension services made available to farmers and increases in awareness, knowledge, adoption, and productivity.

While the AKAP sequence has a natural ordering, it is clear that real resources in the form of skills and activities by both extension staff and farmers are required to move along the sequence. Awareness is not knowledge. Knowledge requires awareness, experience, observation, and the critical ability to evaluate data and evidence. Knowledge leads to adoption, but adoption is not productivity. Productivity depends not only on the adoption of technically efficient practices, but of allocatively efficient practices as well. Productivity also depends on the infrastructure of the community and on market institutions.

Extension services affect each part of the sequence. They can be seen as both substitutes for and complements to the acquired skills of their clientele farmers. Empirical evidence indicates that they are, on balance, net substitutes for farmers' skills as reflected in farmers' schooling. For example, extension services are typically not the only sources of information.

Skilled farmers can seek information on their own. Farmers with few skills may not do so. Extension information then may have a higher impact on farmers with less schooling. It appears, however, that the awareness-knowledge part of the sequence is where extension services are strong substitutes for farmer schooling. Through organised frequent contact, they "teach" farmers, and this is more than simply informing farmers.

The teaching versus informing distinction is also relevant to the "newness" of the information and of the nature of the practice or new technology. When technology is new and is also "simple" to evaluate and adopt, information-awareness is relatively easily converted to knowledge and adoption. Farmers with few skills usually adopt such technology with a time lag. When the technological practice is more complex and requires substantial changes in activities and sometimes capital investment, teaching is required. Repeated messages clearly stated, followed up by field staff and often community organisation, are required to proceed through the AKAP sequence in this case.

DEVELOPING AND DELIVERING EXTENSION PROGRAMMES

Extension planners throughout the world face the difficult challenges of being creative in their programme development efforts and responsive to the needs of rural communities and farmers. A way to overcome these challenges is to look at different programme development ideas and approaches, analysing how they function in practice, as well as their advantages and disadvantages. It is important to emphasise that, although in the text we usually use the word "planning,"on programme development globally, which we see as a set of dynamic cycles necessarily implying planning, implementation, and evaluation stages and activities, interacting and often overlapping, evolving along the programme process and according to the changing circumstances of the physical, socioeconomic, and political-institutional environment. In today's world, the pace of change is accelerated, and people are continually involved in it, either as passive elements or as active citizens, more often as mere project recipients or targets. Development projects are delineated to help people adjust to change, for example to new agricultural policies or market demands. Also, at times, but not as frequently, they are directed to help grass-roots groups and rural communities to build change projects that are relevant to their own needs and aspirations. Both situations require planning and the preparation of different types of extension programmes.

Programme Development

In one way or another, when we talk of programme development, we talk of some kind of planned change, that is, deliberate efforts to change a given state of affairs. Social, economic, cultural, or technological changes are commonly assumed to be the purposes of planned and systematic extension actions.

As change facilitators, extensionists should then be concerned with the preparation of programmes and projects that are responsive to the needs and interests of rural communities and farm families. Some would argue that planning is a difficult mission, especially in the developing world, where the level of political and economic uncertainty is high. This leads to on-the-spot decisions or very short-term and incipient planning work.

However, we strongly argue that the lack of adequate planning and continual evaluation is a major reason for the frequent failure of development projects and extension activities: planners do not look at the diversity embedded in most situations; different clientele groups are not systematically involved, and some (surprisingly, major ones) are neglected; alternative solutions are not carefully compared; objectives are too rigid, not clearly defined, or not linked to activities; results are not duly studied; and the distribution of benefits is overlooked.

So in many situations the challenge seems obvious: to allocate more time for planning and evaluation and to stop acting hastily and mechanically, without direction and purpose, like a clock lacking hands. Extension authors and professionals strongly support planning. Forest and Baker, for instance, underline that "program planning helps justify budget appropriations and brings understanding among the public," adding that "the planning process offers opportunity to people who participate in it to learn, thus building leaders-hip skills in the community that will likely contribute to self-help, independence, and positive end results." In general, adequate planning and evaluation do the following:

- Involve an integrated analysis of needs and interests, opening up new horizons for action
- Promote a concentration of efforts, channelling energies and resources in appropriate directions, and helping accomplish complementarities and synergies
- Strengthen programme resources and attract funds, thus allowing the sustainability or expansion of activities
- Improve team and community capacities, motivations, performance, and autonomy
- Show commitment to address and solve problems
- Strengthen the quality of projects and staff performance
- Serve as a means to open dialogue with other organisations involved in development

In this sense, it is understandable that "planning extension programs has become an increasingly accepted practice among national authorities".

The world has a complex and uncertain nature. Change and the unexpected live with us and are important ingredients in the history of both humankind and the local society. Planning, then, is a risk-taking exercise, subject to the unexpected and to failure. Without it, however, life is even more uncertain and the task of reaching a balanced and sustainable development more difficult. So when talking about extension programme development, we assume that:

– Planned change may be an important factor for the social and economic progress of rural communities and families
– Extension services and agents must not act mechanically and without a vision
– Extension programmes can contribute significantly to learning, educational improvement, and development
– It is possible to select, organise, and manage programmes that contribute to change and development
– Extension educators, as change facilitators, can help individuals, families, and communities to reflect upon their realities and build relevant programmes, thus improving the quality of rural life

Comparing Major Approaches

An ideal or universal programme development model or approach does not exist. Different options are available, and choices depend upon a great number of variables. It is necessary, for example, to have a clear definition of who plans, for whom, who takes the initiative, what the goals are, what the means are, what the time frame is, and what the sociopolitical environment is. Other variables may also be very important, such as the source of funds or the project's physical scope.

It is current in our days to contrast different programme development approaches by using the following dichotomies: centralised—decentralised, top-down—bot-tom-up, and blueprint—process. The two extremes represent, indeed, distinct ways of approaching programme development and correspond to somewhat opposite assumptions, theories, and practices. Dichotomies, however, are simple ways of representing an issue, and in between the two extremes may lie a continuum of possibilities which must not be disregarded.

The first word in each of the pairs is centralised, top-down, and blueprint. In general terms, they correspond to the so-called conventional way of developing a programme. This is, in effect, what happens in many extension projects following the training- and visit system or other conventional model and stressing the transfer of technology and information dissemination. For example, research stations develop technologies which are then transferred to the extension service through subject-matter specialists. Extension officers at the zone or district level plan the programmes, defining specific objectives and messages to be disseminated. At the village level, extension workers implement the activities according to fixed work schedules, under close supervision and leadership. Farmers' involvement is not, in general, a priority.

This type of approach is based on a number of key assumptions and principles: there are clearly defined and generally accepted objectives; there is a detailed and precise knowledge of the process to be implemented in order to reach the objectives; there is the political will to use the available power and resources; and there is a predetermined timetable and well-known resources.

It is generally accepted that the centralised, top-down, or blueprint approach facilitates management, monitoring, and evaluation tasks because activities and expected outcomes are defined and a chain of responsibilities and duties is well identified. In some cases, it can be the best choice, for example, in emergency interventions where a strong management style may be required to attain objectives in a timely and highly organised manner, or in situations where extension tasks are objective, that is, based on specific facts and knowledge rather than on feelings, beliefs, and values. In this last instance, as Boyle stressed, "The programmer would be able to make objective decisions and probably have less need for clientele input."

Top-down or blueprint planning has been subject to strong criticism for various reasons. An important one is that it is too uniform, not taking into due account the sociocultural environment, the particular circumstances in which project implementation occurs, and the characteristics of the different clientele groups, for instance, planning the improvement of village irrigation schemes without taking into consideration the specific local needs, water rights, and rules of water allocation and distribution; or planning for the dissemination of a given technology package without an adequate understanding of the farming systems and the diversity of farmer's problems, potentials, rationales, and strategies. Others also argue the following:

- This approach is agency centred and the programmes are planned from the inside to the outside; planners assess and define needs and problems and determine objectives and courses of action.
- Programmes are essentially based upon institutional policies and philosophy, not taking into account the diversity of perspectives about a given reality.
- The approach assumes a high degree of simplicity and order in the programme cycle, stressing the possibility of following a logical sequence of steps.

- The approach is rigid and assumes a high level of stability; problems will not change while the programme is being planned and new problems will not emerge.

Looking at the second word in each of the above pairs, we have a different set of key words: decentralised, bottom-up, and process. They correspond, in general terms, to what has been called participatory planning, currently proposed as a key element in farming systems development, farmer-first models, participatory technology development, or local process facilitation activities.

The guiding principles are quite different from those of the top-down perspective:

- Development is regarded as a long-term effort and process requiring continued commitment and collective responsibility.
- Programme personnel should act as partners and facilitators rather than experts.
- Participation of local actors is stressed.
- More time should be spent on needs identification and project preparation, with the active involvement of the intended beneficiaries.
- The programmes should grow step by step, securing close linkages to the felt needs and the local environment.
- The ultimate goal of the programme is to increase the power of the local actors to plan and implement their own improvements.

In general, this approach is said to be open and process centred, embracing error as a learning factor and leading to programmes and projects with an emergent nature.

The use of such principles and ideas is growing in small, local development projects with an integrated nature where citizen participation is highly valued and desired.

It is clear that this approach is not exempt from problems and criticisms. Some are quite evident; for instance, activities start without predefined objectives, making things more difficult for personnel and perhaps confusing for participants who often used to be recipients of interventions, not active partners; and the overall philosophy and practices contrast with the conventional ones with which most funders and official services are familiar, which may complicate relationships.

The question of decentralisation is critical in this last respect. In effect, even considering the efforts developed in the last decades in this direction, it is known that the capacities at the local level are still generally weak and "finding ways of building the capacity of local administrative units to implement development programs and of eliciting the support of central bureaucracies in that task offers an important challenge to governments of developing countries in the years to come". Also important is the issue of participation. It can be argued that participation is not desirable in every situation and that it has advantages and disadvantages. For instance, it requires time, which in some instances is not available, and it may lead to social and political conflicts because it touches the question of power, and those who may lose it are likely to oppose and resist. Undoubtedly, each project situation requires a careful analysis regarding the purposes which might be accomplished through citizen participation, the ways to achieve it, and the costs and benefits deriving from it.

In most cases the choice is not at one extreme of the continuum, but somewhere in between. At the same time, a combination of different approaches may even be necessary and advantageous. Boyle, distinguished three types of programmes: developmental, informational, and institutional. The first one is clearly process centred, and objectives derive primarily out of the needs and problems of the participants. The second is predominantly top-down oriented, and participants are primarly recipients of information. The third type corresponds basically to training and instruction and is developed from a field of knowledge and from the educator. However, Boyle stresses that in the course of extension work these types usually overlap; that is, a locally initiated participatory project can very well benefit from centralised efforts to disseminate information and from training activities.

CREATING EXTENSION PROGRAMMES WITH THE PEOPLE

Planning is currently recognised as a process and a social practice. As a process, it is seen as a dynamic effort evolving around problems and implying decisions and actions to achieve goals. As a social practice, it implies a negotiation of interests and the construction of some type of working platform involving different actors, such as researchers, subject-matter specialists, rural leaders, representatives of farmers' associations and groups, and so forth. "Negotiation" emerges as a key word in current views.

Democratic planning of extension work implies the open discussion of these questions, not necessarily in any exact order, but usually in some kind of iterative way, going back and forth in successive approximations. Participation of men and women farmers, rural leaders and networks (both formal and informal), community groups, and other development institutions becomes a critical issue requiring creative approaches and continual efforts.

In fact, as experience shows, the majority of farming and rural populations, often representing the largest share of production, although in words considered central to development, are not actively and systematically brought into the process. It is known that extension work overlooks women and is mostly dedicated to a minority of progressive and large farmers who tend to be relatively well-off.

These farmers have more power to influence the system, serve many times as privileged linking elements between the village and government agencies, are actively involved in organisations, and are capable of attracting a good share of project benefits. On the other side, it is also known that often small and medium farmers lack the time, money, and motivation to participate. They feel powerless, and very often local history is not encouraging and supports the view that participation is politically risky.

Creating programmes with the people implies a broad view of people's involvement. Participation should not be seen, as many times in the past, as the occasional presence of rural dwellers in an information meeting, the simple use of public services, the voluntary contribution (with labour, money, etc.) to a project, or as some kind of activity to increase support to preplanned top-down projects.

Participation becomes, then, people's involvement in reflection and action, a process of empowerment and active involvement in decision making throughout a programme, and access and control over resources and institutions.

A new view about the people and their roles is also essential. In fact, the view that planners, administrators, and others often hold about rural people, small and poor farmers in particular, is very influential. Still relatively-widespread is the idea that local society is practically undifferentiated and that farmers are passive and ignorant and lack the ability to understand situations, analyse them and proposed solutions, translate them into action, and evaluate the results. The direct consequence is the paternalistic and superior attitude of technicians who think they know best and have the right answers to development problems.

REFERENCES

Jones, G.E. and Garforth, C. 1997. "The history, development, and future of agricultural extension" in Swanson, B. *Improving Agricultural Extension: A Reference Manual.* 3rd Edition." FAO.

Leeuwis, C. and van den Ban, A. *Communication for Rural Innovation: Rethinking Agricultural Extension* 3rd Edition., Blackwell Publishing.

Nagel, U. J. 1997. "Alternative Approaches to Organizing Extension", in Swanson, B. *Improving Agricultural Extension: A Reference Manual.* 3rd Edition." FAO.

Roling, N. 1988. *Extension Science: Information Systems in Agricultural Development,* Cambridge University Press.

Trager, J. 1996. *The Food Chronology,* Aurum Press, London.

3

Worldwide Extension Programmes

Extension, in general terms, is a function that can be applied to various areas of society. It operates in the industrial, health and education sectors, as well as agricultural and rural development. Originally derived from "university extension", the term "extension" is therefore applicable to various areas of development.

Extension functions in various sectors of society. The World Bank-sponsored Training and Visit (T&V) extension system, Israel stressed the fact that T&V principles could be applied to other sectoral systems involved in the delivery of nonformal education. It also warned, however, that T&V was based on classical management principles that were unlikely to be viable in the developing countries—a lesson that was only fully appreciated in the 1990s and that has since led in part to the current emphasis on participatory management principles.

AGRICULTURAL EXTENSION—KNOWLEDGE SYSTEM

Agricultural extension operates within a broader knowledge system that includes research and agricultural education. FAO and the World Bank refer to this larger system as AKIS/RD (Agricultural Knowledge and Information Systems for Rural Development). The OECD countries refer to it simply as the Agricultural Knowledge System (AKS). Others describe the three pillars of this system—research, extension and agricultural higher education—as "the agricultural knowledge triangle" and suggest that since the three pillars involve complementary investments they should be planned and sequenced as a system rather than as separate entities.

At the second OECD AKS conference in January 2001, the representatives of agricultural research, education and extension institutions, and government officials with AKS policy responsibilities stressed the opportunities for AKS to address the wider societal issues associated with agriculture.

The participants confirmed that AKS could play a central role in developing research, education and extension/development programmes oriented towards these wider society issues, which can be expected to expand. In this scenario, the planning and sequencing of AKS as a single system, becomes even more imperative. However, much has been written on implementing AKIS linkages, especially in research and extension without any significant results.

According to the AKIS/RD Strategic vision and guiding principles the system integrates farmers, agricultural educators, researchers and extensionists, enabling them to harness knowledge and information from various sources to improve farming and livelihoods. Maguire suggests that the concept and practice of agricultural education should be redesigned in the developing countries as education for rural development and food security.

Indeed, many needs are rapidly emerging such as trade-related education on agro-health, value-added agro-processing, and agro-market competitiveness. These needs arise from the obligations that countries take on as members of the World Trade Organisation (WTO) and the increasing urgency to build competitive advantages aimed at global agricultural market niche opportunities. In principle, agricultural extension receives relevant information from the agricultural education system and feeds back field observations to this system.

Extension is also professionally linked to the agricultural vocational and higher education systems in the sense that these systems also produce the agents who work in extension. The relationship between agricultural extension and agricultural research is even closer, because the knowledge that agricultural extension transfers is usually generated by agricultural research through applied and adaptive agricultural research development. Within the agricultural sector, however, agricultural extension may be interpreted narrowly or broadly, which complicates the debate.

In a strict interpretation, the only purpose of agricultural extension is to disseminate information to raise the production and profitability of the farmers (agricultural production performance). In a broader interpretation, the purpose of agricultural extension is to advance not alone production knowledge but the whole range of agricultural development tasks, such as credit, supplies, marketing and markets (agricultural process development).

In the broadest interpretation, agricultural extension provides nonformal—agriculturally related continuing adult education—for multiple audiences: farmers, spouses, youth, community, urban horticulturalists (continuing agricultural education and community development) and for various purposes.

In some countries all three of the above orientations operate, e.g. the U.S. Cooperative Extension System. Such extension systems encourage the empowerment of farmers in various ways, including participation in programme planning and decision-making. By contrast, in many countries agricultural extension is linked to agricultural production services.

CONCEPT OF AGRICULTURAL AND RURAL EXTENSION

When agricultural extension is combined with rural extension goals, the extension function ranges even more widely in its purposes. Rural extension, for instance, includes non-agricultural activities such as microenterprise development, a priority which is being advanced by the Inter-American Development Bank. Non-farm rural microenterprise development. Most rural people depend upon multiple sources of income, such as petty trade, primary production, remittances, and casual employment.

In short, rural people are not dependent solely on agriculture or natural resources for their livelihoods. As Carney points out, "these might provide the basis for their survival but it may well be that the best prospects for significant livelihood improvement lie outside the natural resources sector in the generation of off-farm income". In addition to microenterprise development there is also the option of reaching the poor through rural public employment, i.e., labour-intensive rural public works projects.

Since the AKIS/RD document combines rural with agricultural goals, and since rural development involves both farm-related and non-farm-related activities, it seems appropriate for certain extension programmes to be engaged in activities beyond those already mentioned. FAO could promote the development of agriculture- related micro-enterprises in rural areas where such a priority would make sense for extension programmes, and in this regard it might launch a special alliance with relevant organisations such as the Inter-American Development Bank.

Technical Extension

Agricultural and rural extension is the responsibility of various technical and service units, and serves many purposes. The various technical units within FAO indicate that agricultural extension is a function pursuing many different purposes: livestock development, forest use and conservation, fisheries engineering and capture, food and nutrition education, as well as well as crop development. Even in programmes designed to foster agricultural crop production, extension may be concerned with providing information on other crucial issues such as food storage development, processing, farm management, and marketing. FAO has advocated and pursued all the above purposes of agricultural and rural extension at some time or another.

Marketing Extension

Other purposes of agricultural and rural extension include marketing extension. Marketing extension provides information on the post-harvest treatment of speciality crops and provides an important service in countries trading in food crops, including such fragile products such as bananas and cacao. Other, different types of marketing information services referred to as "market extension" also exist; these services provide information on variations in commodity prices; knowledge about where to sell some products; information on problems to do with the quality, availability and prices of inputs, and on the actual level of competition in the markets. These market information services should not be confused with marketing extension services that aim at improving the preparation and process of moving agricultural goods to market.

Farmers' Associations

Agricultural and rural extension services can also help farmers and produce processors to organise themselves to meet their mutual agricultural interests. A long tradition in extension is group promotion and group organisation, and FAO's commitment to these purposes is well known. Indeed, one of the Organisation's many ways of promoting people's participation in development is through independent agricultural and rural development group associations. Financing economic selfreliance and the participation of the members in their organisation's activities is of central importance in such efforts to promote farmers' organisations.

Some argue that extension can most effectively carry out its mandate, not by working directly with individual farmers but by working indirectly with and through farmers' groups or organisations.

Emerging Purposes

As populations grow and rural peoples flock to the cities, extension may have to deal with urban and suburban clients. What is currently considered "agricultural and rural extension" may eventually become "food and agriculture, rural and urban extension ". In fact, extension in high-income countries is already providing information and education services in urban areas, extending beyond technical agriculture and rural development alone. Urban extension is a potential growth area for information transfer. As such, it addresses new audiences and new programmes, and reflects the world's rapid urbanisation.

In Latin America, for instance, urbanisation will affect 83 percent of the population by the year 2020. This process will involve socio-economic and demographic changes that will affect food and nutrition, as well as epidemiological, institutional and sociodemographic changes. The same process is apparent in Asia and Africa, as well as in North America and Western Europe. Food security, the employability of youth in the food industry, environmentally sound practices by small urban businesses, and other food and agriculture-related programmes are likely to demand the attention of governments which are currently dismantling extension programmes. Conceiving of extension purely as an agricultural production, rather than an educational service is short-sighted and limited.

ALTERNATIVE EXTENSION APPROACHES

The Extension, Education and Communication Service (SDRE) issued a valuable overview of extension approaches and methodologies in 1988 authored by George Axinn and entitled Guide on Alternative Extension Approaches. This overview is still a valid reference work and provides a basic examination of the various extension approaches current at that time. The guide distinguishes between eight different approaches.

The terms "approach" and "methodology" when referring to extension are often used interchangeably, and it would be pedantic to try to separate them and their respective conceptual meanings. Suffice it to say that while the terms may be used interchangeably, they must be differentiated from the term "options" used in refer to reform strategies involving institutional arrangements. Most of the approaches to which Axinn refers have been supported by FAO at various times. These eight main approaches are simply listed below, for sake of brevity, together with their respective success criteria.

1) *The general agricultural extension approach.* Success is measured in terms of the rate of take-up of the recommendations, and increases in national production.
2) *The commodity specialised approach.* The measure of success is usually the total production of the particular crop.
3) *The training and visit approach.* Success is measured in terms of production increases of the particular crops covered by the programme.
4) *The agricultural extension participatory approach.* Success is measured by the numbers of farmers actively participating and benefiting, and the continuity of local extension organisations.
5) *The project approach.* Short-run change is the measure of success.
6) *The farming systems development approach.* Success is measured by the extent to which farming people adopt the technologies developed by the programme and continue using them over time.
7) *The cost sharing approach.* Success is measured in terms of farm people's willingness and ability to share some of the cost, either individually or through their local government units.

8) *The educational institution approach.* The measure of success is the farming people's attendance at and participation in the school's agricultural extension activities.

This is certainly not intended to be an exhaustive type listing. More importantly, Axinn's characterisation of the different approaches in terms of their success tends to distort some of them. It nevertheless helps to distinguish certain basic approaches.

Why is there such a plethora of extension approaches? Some ideas change; paradigms shift; and purposes vary. But lessons are also learned, and then shared. It becomes clearer why one or other approach has succeeded or failed, and which aspects of a particular programme are useful and which are not. Even a cursory review of FAO's agricultural and rural extension approaches indicates the diversity of its involvement. Clearly, agricultural extension involves many different approaches and methodologies. It is also directed towards very distinct content areas. And it is managed and delivered through a variety of institutional arrangements. It can therefore reasonably be argued that no single approach best suits extension development in all circumstances, just as there is no one single approach that best suits development. Otherwise the problems of extension and, for that matter, of development, would have been solved long ago.

GOVERNMENT'S ROLE

Government plays an important role in agricultural and rural development, although its relationship to extension funding and delivery is changing. Even when agricultural extension is farmer-led, government—at whatever level—must be concerned with production, the impact of agricultural practices on the environment, regulations governing quality standards, food safety, and in general the well-being of the people. There has arisen a myth about "the powerless state".

However, it is no myth that government extension has in many cases become irrelevant and has been by-passed by NGOs and private commercial extension. In the final analysis, though, it is government that decides whether or not to become directly involved in agricultural and rural extension.

Governments are facing new extension challenges: meeting the need to provide food for all, raising rural incomes and reducing poverty, and sustainably managing natural resources. These critical challenges exist in a rapidly changing world. Globalisation, new technologies, the new relationships developing between the public and private sectors, the multi-disciplinary nature of agriculture, heterogeneity between and within countries, the geographic dispersion of rural people—all these realities are putting new pressure on the developing countries in their efforts to develop. This being so, the state must take on a central role in financing advisory services which are important, but not financially rewarding for the private sector.

In addition to providing advice on the management of natural resources, integrated pest management and advisory services to the very poor, the state has a critical role to play in establishing markets for commercial and farmer-to-farmer extension services, providing rural communication infrastructure, and developing human resources.

The advancement of pluralistic partnerships is crucial, given the multiplicity of tasks confronting developing countries. For those governments that have not yet done so, the advantages and disadvantages of institutional reform deserve consideration. In this regard, governments as well as international organisations need to benchmark the pros and cons of newly reformed institutional arrangements for agricultural and rural extension systems, and learn from each other.

Institutional reforms appear to have been successfully carried through in various countries and may be of value to governments when considering the possibility of reforming their own agricultural and rural extension. However, no single reform measure can be considered a panacea. All are "work in progress" and depend on the commitment, resources, capacity, attitudes and motivations of the stakeholders at various levels.

GLOBAL DEVELOPMENTS SHAPING EXTENSION

The role of agricultural extension is vital to the diffusion of new technology, but extension is currently failing or moribund in many African nations. In other low-income developing countries, extension is in disarray or barely functioning at all. Staff are bloated, under-trained, not mobile, and therefore not proactive. There is also little, if any, coordination between extension and research, and even less between extension and agricultural higher education. Meanwhile, other forces are affecting the development of agricultural and rural extension services emphasizing issues and challenges for much needed reforms.

A "complex" of extension providers has emerged, involving nonprofit non-governmental organisations and for-profit private companies, and farmer organisations and commercial associations of extension specialists have also come to the fore. In some cases, these non-public sector extension service providers hire public sector extension agents on secondment, which has been termed "contracting-in". The world's expansion beyond the global village is a reality that has strongly affected public sector extension.

Globalisation is inextricably linked to privatisation, and countries are finding themselves confronted by a new and highly competitive global market. Major economic restructuring is taking place in both the developed and the developing countries, and has greatly changed the balance of responsibility between the public and private sectors. In many cases, trade liberalisation places the developing countries at a disadvantage in the global market.

Increasingly privatised, agricultural information has in fact become a price-tag "commodity". This 'commodification' of agricultural knowledge is a major factor in the present worldwide transformation of public sector agricultural extension and the advancement of private sector technology transfer systems.

This change towards information commodification reflects the privatisation of information and agricultural industrialisation. One result is that farmers, especially in high-income and middle-income countries, have begun to pay for extension services. While this may not yet be the case in many lowincome countries, the trend indicates the value placed on agricultural information. One way of reducing poverty is to generate incomes through the training and information-sharing that agricultural and rural extension services can provide. It therefore appears to be in the interests of low-income countries to promote services to provide practical, income-generating agricultural information to their rural populations.

Apart from the immediate value of educating their rural populations in the financial value of agricultural information, they will also help them to reach a financial level at which they can pay for such services. Contemporary extension reform strategies demand attention because they have long-term consequences on economic growth and short-term consequences in terms of improving rural situations by helping to reduce poverty.

THE NEW PARADIGM

The world has in many ways become smaller. Globalisation has occurred with the ease and rapidity of the development of transport and telecommunications. There is a tendency towards greater transnational corporate development. Some argue that there has been a "power shift" from public sector dominance to private sector hegemony.

As this development emerged, fiscal problems in the public sectors of the developing countries became apparent, and donor organisations began to impose structural adjustment programmes to bring the developing countries into line with the financial demands of a stricter financial "world order". These structural adjustment programmes have strongly impacted the national governments of less developed countries—many of whom were, and continue to be, under pressure to reform their public sector systems. A new paradigm towards market-driven reforms and with an agribusiness orientation has resulted from this, severely affecting the funding and delivery of agricultural and rural extension. These changes have produced a turning-point, and are having radically repercussions in term of the way public sector agricultural extension is conceived and practised.

The agricultural sector faces increased competitive challenges. Urban centres will continue to attract people from rural areas. The advancement of science and technology will increasingly pressure countries to modernise. Technologies must be tailored to new contexts if they are to be effective, and such adaptations require an educated workforce. Preparing an educated workforce will require considerable investment in education, in-service job training (e.g. for under-trained agricultural extension and rural development agents), and the knowledge exchange component of the technology system.

At the same time, governments will probably find it difficult to continue to operate status quo ante. New strategies are taking on primary importance for policy-makers concerned with meeting the demands of the global market place, while at the same time catering for the needs of their rural populations.

Lowincome countries are faced with the inescapable challenge of qualitatively and quantitatively increasing the rate at which they are moving forward if they are to survive in the global market place. Yet it is still obvious that there is under-investment in extension, and in research and agricultural education.

In some cases, the solution to the developmental problem may be not to improve agriculture. Where there is little potential, an agricultural programme cannot have much impact and there will be only small returns on investment. Out-migration in the short-term, or industrialisation in the long-term, may be the appropriate solutions. Where agriculture is a viable development option, commitment should be clear-cut, recognising that there will be different target groups to be served. The implication of target group analysis is that there must necessarily be different extension systems to meet a wide variety of different needs.

Extension or extension- related services will become more purpose-specific, target-specific, and need-specific. In some instances, such services will transmit high-tech messages. In others, farm management skills will be developed. In some instances, farmers will organise themselves for marketing purposes. In short, a wide range of institutions will be established and developments take place, expanding the range of educational, organisational and technology exchange services for agricultural and rural development.

Responding to the New Paradigm

At this juncture, farmers have to be convinced that extension systems and the information they communicate are valuable for income-generation in particular and for improving their living standards in general. Assisting resource-poor farmers with appropriate technology may provide the opportunity for rural households to increase their productivity and incomes. In some cases, the new opportunities envisaged by farmers may slow down rural-urban migration.

Small, low-resource farmers represent a vast segment of developing country populations; higher incomes, education, and greater involvement in development can also encourage them to make more efficient use of the land, labour, and capital resources in rural areas. This can be done in several ways, according to Swanson; for instance,

(a) small farm households could be helped to intensify and diversify their farming systems,

(b) small farmers need to be brought into the market economy,

(c) small farmers also need encouragement to practise agricultural sustainability, and

(d) small farmers need to be helped to organise themselves around their mutual agricultural interests.

Swanson's language may appear supply-driven and top-down, but the basic aims expressed are in line with current development needs.

There is a growing consensus that to create a demand-driven technology system there must be direct involvement by farmers in identifying problems, establishing priorities, and carrying out on-farm research and extension activities.

Demand-driven extension is desirable in many instances, although a balance must be struck between the demands of government and those of farmers in different economic categories. Striking a balance between institutional "supply systems" and farmer-initiated demand-driven extension/technology systems should in many cases be the ultimate goal of countries eager to advance to higher stages of development and competitive power.

FAO'S PROGRAMMES

FAO makes a major contribution to developing agricultural and rural extension institutions and programmes. It is actively committed to fostering multiple extension approaches and methodologies, organising world conferences, exchanging information with external sources, and undertaking many field activities. Underlying this diversity, however, is FAO's special concern for participatory approaches and the development of equitable processes. It is this emphasis on participation and people-centred approaches that distinguishes FAO, and forms a central concern in its promotion of agricultural and rural extension.

Multiple Extension Approaches and Purposes

FAO's assistance in support of agricultural and rural extension tends to involve various approaches and distinct methodologies for interacting with the clientèle and for interchanging information. One outstanding FAO participatory approach is known as Farmer Field Schools. Farmer Field Schools (FFS), originally associated with promoting Integrated Pest Management, work at the grassroots level to advance the principle of stakeholder participation in programme decision-making with a view to eventually giving full responsibility to stakeholders for programme development.

FFS underscores FAO's commitment to the development of agricultural extension participatory approaches—in line with its general philosophy and practice of seeking to advance equitable development. Originating in projects initiated in Asia in the mid-1980s, FFS has spread to other regions. And today, FFS is beginning to develop in Latin America as one of the alternatives to traditional national extension activities, in such countries as Bolivia, Ecuador and Peru.

In its efforts to promote farmer-led extension FFS is proving to be a viable alternative to centralised and state-owned extension. The approach is currently one of the forefront extension- related activities sponsored by FAO, and the principles and methodology of the approach are being replicated by other technical services such as Irrigation and Water Use and Forestry. In fact, the Irrigation and Water Use technical unit has already successfully piloted an FFS project in Zambia.

"A Note on the Sustainability of the Farmer Field School Approach to Agricultural Extension" provides an interesting perspective on FFS as an alternative learning or problem-solving approach. They view FFS not as an extension approach for disseminating information, but as an empowerment and citizenship opportunity. At the same time, they raise FFS cost issues and their relevance to the sustainability of this approach.

FAO's Forestry Policy and Institutions Branch has also adopted the FFS approach, but has changed the name to suit its community forestry development purposes: Farmers' Forest Management Schools (FFMS). According to Tanaka, "FFMS has two objectives. The one is to allow forest users flexible community forest management for multiple use. FFMS assists forest users to gain/generate the knowledge, critical skills and self-confidence to make decisions about forest management based on their own experiments, observations and analyses so that the forest can sustainably provide them benefits suitable to their livelihood needs.

The other is to provide a platform for negotiation among various forest users in the process of determining intended use of community forest. This process also helps them build the sense of ownership through delegation of decision- making and forest management process". Another participatory programme approach is Farming Systems Development. Farming Systems Development (FSD) began in the 1980s as Farming Systems Research and Development and later became known as Farming Systems Research and Extension.

On-farm research is seen as a link between farmers, technical research and extension. This approach has a dual character. Sometimes it is hailed as a multi-institutional, team approach; at other times it is considered a production-oriented approach. The term 'Farming Systems Research' (FSR) tends to suggest a production-oriented approach, while the term 'Farming Systems Development' (FSD) suggests more of a team-oriented approach involving the farmers in the process. FAO has supported both these system approaches.

FAO's technical units also maintain or utilise extension services for their specialty purposes. For instance, Fisheries is creating distance education tools to extend information for modern fisheries engineering and fisheries capture techniques. This computer-based, distance education programme promotes "learning by doing". Distance learning is not only a major development in information and communications technology (ICT) but is already a leading instrument for extending information and knowledge.

The livestock development unit is also engaged in information exchange, especially for preventive animal health services. Extension methods are employed to reach pastoralists and breeders. Animal health is a major concern, as Europe's current problems aptly indicate. As the list of those services providing a specialised knowledge base and new technology grows, the different purposes for extending information become obvious.

The value and importance of extension for different purposes extend even into areas that do not directly create extension services. Nutrition services, for example, while not engaged in extension per se, cooperate with extension services to respond to food crises in the developing countries and to manage food distribution and education.

Likewise, the Socio-economic and Gender Analysis (SEAGA) programme incorporates extension concerns. As an analytical programme aimed mainly at training development workers with methods and tools for conducting socio-economic and gender analysis, SEAGA seeks to heighten the awareness of gender issues and strengthen the capacity to incorporate gender considerations into development. Adopting a bottom-up approach to identifying development priorities, SEAGA promotes the participation of all the stakeholders, emphasizes gender roles and relations, and includes disadvantaged people as one of its priorities.

An interdisciplinary programme, the Integrated Support to Sustainable Development and Food Security Programme, has been set up to develop and implement an integrated strategy that takes into account the main social, economic, environmental and technical aspects of sustainable development.

One of the seven areas addressed by IP is the reform of agricultural extension systems to support poverty reduction, the integration of gender issues and sustainable development. Apart from the technical extension programmes just mentioned, the Extension, Education and Communication Service (SDRE), which is the lead technical unit responsible for agricultural extension and training, education, rural youth, and communication for development, has embarked on several innovative programmes, including the National Agricultural Extension Systems Reform Initiative (NAESRI). Initial work under NAESRI includes strengthening decentralised extension services, developing the coordination mechanism for pluralistic extension delivery, conduct best practices studies, draw up participatory extension delivery strategies for female extension workers, identifying the extension service needs of physically disabled farmers, developing of extension strategies to deal with HIV/AIDS, and encouraging the wider adoption of hybrid rice cultivation.

A comprehensive extension and production support strategy, and a participatory farmer group extension approach have also been developed in support of the FAO Special Programme for Food Security (SPFS) in Pakistan and Tanzania, respectively. Training modules on integrating the population and environmental messages into ongoing extension programmes have been prepared in Bangladesh, Egypt, Indonesia, Nepal, The Philippines, and Thailand. Monitoring and evaluation guidelines have been developed for national extension systems in the Caribbean. A number of country studies on AKIS/RD have been initiated in collaboration with SDRE's sister unit, the Research and Technology Development Service (SDRR). SDRE has also developed the Strategic Extension Campaign (SEC) methodology, which has been implemented in a number of developing countries.

SDRE, with SDRR and the World Agricultural Information Centre (WAICENT), has initiated FarmNet, a community-based programme that aims at creating farmer information networks for agricultural and rural development. This programme uses communication processes and tools to facilitate the generation, gathering and exchange of knowledge and information among rural people, and between them and the intermediary organisations that work for them.

The Uganda National Farmers' Association is one of the organisations that have shown interest in the programme. SDRE and WAICENT are jointly developing another programme, VERCON (Virtual Extension and Research Communication Network). Unlike FarmNet, VERCON is institution-based and aims at promoting communications between research and extension systems. These initiatives seek to exploit the great potential of electronic mass media and telecommunications to improve the transfer and use of agricultural information in the developing countries.

WAICENT and SDRE are developing a new participatory programme on "Information in Support of Sustainable Livelihoods". While still in progress, this sustainable livelihoods programme plans to promote a "mix of media" and various processes, procedures, methods and tools for promoting participatory information exchange between indigenous groups and the external environment. In sum, FAO assists the developing countries in numerous and varied ways with the development of agricultural and rural extension. One common feature of these different endeavours is the participatory involvement of the people being assisted.

All of the above-mentioned efforts—e.g. FFS, FFMS, NAESRI, FarmNet and Sustainable Livelihoods—encourage stakeholder involvement in the extension decision-making processes. Empowering local communities and small farmers in the use and development of extension services using participatory approaches remains one of FAO's most central and important goals. In general, FAO's ground-level programmes underscore a number of goals aimed at fostering rural people's advancement through agricultural and rural extension services.

These goals include (a) improving the skills of farmers, extensionists and extension managers; (b) empowering farmers through participatory extension; (c) assisting special end-users, especially women; (d) cooperating with other agencies and organisations in extension development; and (e) assisting governments in extension policy formulation and programme development. Improving the relevance and effectiveness of agricultural extension activities for women farmers deserves priority status in any consideration of extension institutional and programme development. FAO's programmes reflect the diversity and stress the continuity affirmed in the framework of its Mid-term Plan and implemented through its technical cooperation programmes. The Organisation's current programmes set the stage for what to do next.

NEW VISION FOR AGRICULTURAL EXTENSION

In an effort to respond to the new paradigm, countries worldwide have adopted a variety of institutional reforms. These reforms are either market-oriented or non-market-oriented.

As a prelude, it is worth noting that the types of extension reforms being enacted are not necessarily new. Decentralisation is certainly not new to extension. In some countries extension is historically decentralised, and devolved authority has long existed in Australia, Brazil, Canada, Germany, India and the United States. Partial privatisation has existed in France since the 1960s. Participatory extension methods are a traditional aspect of extension in Japan and Taiwan. Pluralistic systems have a long history in Finland. What is new, however, is the extent of globalisation and, as Fresco notes, the major economic restructuring in both the developed and the developing countries which has greatly tipped the balance between the public and private sectors.

SDRE has been working for several years in various countries, francophone, anglophone and lusophone, to promote such reforms as decentralisation, participation, and provider pluralism. Examples of these efforts in the francophone countries are FAO's work in the Congo, Rwanda, Guinea-Bissau, and the Central African Republic. References to the latter efforts exist only in internal FAO technical documents.

Exploratory efforts towards pluralism have been carried out in Mozambique and Zimbabwe. Subsidiarity has become part of the current drive towards reform in Uganda. The Philippines, Indonesia, Iran and Yemen are other examples of efforts at decentralisation aided by SDRE. Various aspects of these reforms have, in fact, been integrated in these countries, demonstrating where they overlap and their complementarities. Nevertheless, what is new is the growing commitment of FAO to promoting these reforms as measures that other developing countries should consider. SDRE's initiative NAESRI is evidence of this phenomenon.

Contemporary Institutional Reform

Global developments are shaping extension even more radically than other institutions in the agricultural knowledge systems. Contributing to this drive to reform extension is the new paradigm supporting market-driven income-generation. A considerable variety of public sector reform strategies have emerged, and can be categorised in different ways. These extension reforms are organised in this subjects under two main headings: market reforms and non-market reforms.

According to this distinction, market reforms encompass four major reform strategies: revision of public sector extension systems, pluralism, cost recovery, and total privatisation. Non-market reforms comprise two main reform strategies: (a) decentralisation, transferring central government authority to lower tiers of government, and (b) subsidiarity, transferring or delegating responsibility, sometimes by abolishing authority over extension, to "the lowest level of society as is practical and consistent with the overall public good".

Market Reforms

Market reforms result from the central government's aim of privatising the management of agricultural and rural extension systems, whether by contracting the delivery of extension field services, obtaining cost recovery by charging fees for the services, or creating partnerships with farmers' associations. The least radical way of reforming a public sector extension system is revising it by down-sizing and minimum cost recovery for services. This revisionist strategy has taken various forms.

For instance, the United States in the late 1980s shifted away from a discipline-oriented, management-by-objectives approach to an issues-oriented management approach to extension, meanwhile gradually introducing charges for previously free services such as soil sampling and attending workshops. Several other governments maintain agricultural extension units in their ministries of agriculture, e.g. Australia, Canada, Japan, Poland, Portugal, and Spain. These countries continue to provide public sector funding and delivery of extension services, either partially or entirely. Reforms that advance institutional pluralism are being widely promoted, especially in the developing countries today.

This strategy often involves contracting out the delivery of extension field services to non-governmental organisations, such as non-profit NGOs, or to for-profit companies including consultancy firms and farmers' cooperatives. Contracting-out is considered an opportunity for public sector reform and private sector development in the developing countries, including the transition economies of Europe and Central Asia.

Kidd et al. cite a variety of experiences with privatisation and the commercialisation of agricultural extension delivery that involve public sector partnering with other entities, viz., farmers' associations in China contracting technical services from public officials, share-cropping for profit in Ecuador, voucher schemes in Costa Rica, sub-contracting and voucher schemes in Chile, privatised service centres in Ethiopia, contract farming in Kenya, and farmer service centres in Sri Lanka.

Institutional pluralism promoted by central government should not be confused with the fact that in most countries there exists an institutional "complex" of public, private, and semi-public service providers. In the case of institutionalised pluralism, the public sector funds the extension activities while the private sector delivers them. In cooperation with TCI and the World Bank, SDRE has been assisting Mozambique in planning to contract the delivery of extension services in pilot provinces by non-profit NGOs, while continuing with public sector delivery in the rest of the country.

In Zimbabwe, where so many public and private service providers are active in extension, SDRE has recently conducted a study that could help in developing a coordination mechanism and define the role of government in a pluralistic situation. In Western Europe, many agricultural advisory services have redesigned their fiscal arrangements, initiating cost recovery or fee-based services to farmers.

Fiscal redesign through direct charging for extension services first became prevalent in the European OECD member countries, largely because of national deficits resulting in changes in government policy and the consequent re-organisation of their Agricultural Advisory Services. Over half the OECD member countries currently receive at least 20% of their finances from direct charging, and two countries (Finland and Norway) receive more than 50% of their finances from users. With few exceptions, agricultural extension services tend to be sold to users at a nationally-determined price; however, in some cases highly individualised projects command higher prices, while projects for low-income users are offered at reduced prices.

Some countries offer discounts or subsidies or otherwise encourage emerging and low-income farmers to procure group-use services that are less expensive than one-on-one services. In Latin America, Mexico adopted a fee-based strategy in the late 1980s for services to large farmers in the northwest grain-producing states, although Mexico later opted to decentralise national extension services to the state level.

An interesting experiment in "fee-charging extension" has been taking place over the past two decades in China. The Chinese central government appealed to local authorities to enhance extension budgets through feecharging extension (FCE) services. The function of FCE is not to recover costs, as it is in such countries as The Netherlands or the United Kingdom, but to act as an incentive. This incentive mechanism seeks to encourage technicians to go out to fields, contact farmers frequently, and ensure that the technologies are adopted. Technology transfer is carried out on a contract basis and is disseminated in response to the farmers' demands.

Although not feasible in all instances, this system of direct contracting between the extension technician and the farmer distinctly differs from the schemes generally cited.

Farmers and extension technicians are closely associated in this scheme, however—with rights, responsibilities and economic interests linked by contract directly between the farmer and the technician. Such an arrangement necessarily demands high quality technical expertise and training on the part of the extension technician.

Administrative and technical training is crucial for the success of institutional reform. To become extension technicians, agents must acquire high quality technical expertise and training. To meet the challenge of reform measures, managerial staff must develop high quality administrative expertise relevant to arranging and monitoring contracts, as well as supervisory skills in human resource management and programme monitoring and evaluation.

Extension workers need to become agronomic, livestock or fisheries technicians or, depending on the programme goals, to develop special skills in farm management, farmer organisation, or other programmatic and organisational skills. Extensionists must be prepared to respond purposively to the requirements of institutional reform, and new knowledge and skills must be promoted as well as new attitudes towards change. Specifically, the premiss underlying the strategy for fee-charging is that farmers will have access to practical information provided by knowledgeable technicians.

One of the most radical recent extension reform strategies is total privatisation, including commercialisation. In the case of total privatisation, both the funding and the delivery of extension services are shifted entirely, or largely, to the private sector, as in The Netherlands. The Netherlands decided to privatise its public extension agents, at first by transferring them with initial financial support to work with farmer associations, and more recently assigning responsibility for these services to a private company, DLV. Privatisation relieves the government of a fiscal burden, often improving the delivery of services once the private sector has taken over the function, although this strategy may leave poor farmers and rural workers without any support. In the case of commercialisation, authority is given to a government-commercialised public agency, as in New Zealand.

New Zealand's Ministry of Agriculture and Fisheries was challenged to "go commercial" in 1986, and to operate under user-pay commercial criteria. This former public sector service now operates as a company, under the name 'Agriculture New Zealand'. A similar step was taken in England and Wales. The Agricultural Development Advisory Service (ADAS), formerly a Ministry of Agriculture, Fisheries and Food executive agency responsible for providing extension services, was privatised in 1997. This body, now known as ADAS Consulting Ltd, is one of many private agricultural consultancies that offers a range of advisory services to primary producers in agriculture and horticulture.

In short, the reforms cited here are not static but are themselves in a state of flux. Several countries, such as Chile, have revisited their extension system arrangements on several occasions and have modified them each time. Chile has moved away from a voucher system to direct subsidisation under contracts with farmers to hire private extension service providers. In most cases, extension reform would best be described as "work in progress".

Non-market Reforms

While central government market reforms aim at privatising—wholly or partially, directly or indirectly—the management of agricultural and rural extension systems, non-market reforms aim at relieving central government of the responsibility of funding and managing extension. The most common non-market reform strategies are decentralisation to lower tiers of government, and transferring or delegating the responsibility for extension to non-governmental organisations, or removing government responsibility for extension entirely.

Decentralisation is often narrowly defined to signify shifting authority for extension to lower tiers of government. Some use the term, "devolution " for this type of political decision. Whatever the terminology, decentralisation is a policy determination that differs from market reforms. Decentralisation reforms are observable in Colombia where extension is being transferred to the municipal level, or in Mexico where authority for the service has been shifted to the states, as well as in other Latin American countries.

Some countries are in the transition stage of moving towards a regionally decentralised extension system, as in Malawi. In collaboration with the TCI and the

World Bank, SDRE has been assisting Uganda to plan shifting administrative and fiscal authority for extension to local government and farmers' groups. Similarly, Yemen has been given assistance to implement its plans to give farmers wider extension responsibilities, as have Indonesia, Iran and the Philippines to help them establish and/or strengthen their devolved extension services. Economists tend to concentrate on the intergovernmental transfer of powers and responsibilities. They argue that the decentralisation of powers to local government is unlikely to be a panacea to the shortcomings of a weak central government.

Governments also "decentralise" to other organisations and levels. At least three decentralisation directions currently dominate the development of agricultural and rural extension. One is to decentralise the burden of extension costs by redesigning the fiscal system. Another is to decentralise central government responsibility for extension through structural reform.

The third is to decentralise programme management through farmer participatory involvement in decision-making and, ultimately, helping farmers take responsibility for extension programmes. Governments may shift authority to farmers' associations, such as cooperatives or chambres d'agriculture, as is being done in Denmark, Finland and France, in Europe, and Ecuador in Latin America. In some cases, governments may shift authority to NGOs or farmers' groups upon terminating government responsibility for public sector extension services. In Latin America where non-market reforms involve delegating authority to private NGOs, as in Bolivia, Ecuador and Peru, this type of reform has been referred to as "tertiarisation". The last form of "decentralisation" is referred to as "subsidiarity". Subsidiarity is a comparatively recent term which, according to the AKIS/RD Vision and principles, means that authority should go to the lowest level at which economies of scale and scope are not compromised and all costs and benefits are internalised.

Giving local communities responsibility for programmes is an attractive option because it integrates local government and rural people into programme design and implementation. Some governments have shifted the institutional and technical responsibility for exchanging and transferring information to the farmers, who manage the agricultural extension programmes themselves. Such participatory involvement is thought to make services more responsive to local conditions, more accountable, more effective and more sustainable. This form of "subsidiarity" exists in Bolivia and Peru. As a rule, subsidiarity highlights government delegation of authority to nongovernmental entities, and in some cases the total disengagement of government from agricultural and rural development interests, giving NGOs or farmers and farmers' organisations that responsibility. In several European countries, such as Denmark and Finland and partially also France and Germany, the farmers' associations carry out extension with partial government sponsorship.

Some countries have undertaken reform through the 'deconcentration' of authority to lower branches of central government. This effort may be a stepping-stone to later decentralisation. At the moment, central authority structures may be deconcentrated to field levels in a number of ways: through financial grants, local coordination, district administration, provincial development planning, and regional coordination, as occurs in Belgium, England and Wales, Indonesia, and Ireland.

Another form of extension development is when dual authority structures are established, with power shared either (i) between the government and farmers' associations, as in Norway and Sweden, or (ii) between the government and a subnational governmental entity, as in Japan, South Korea, and Taiwan. In short, countries worldwide have adopted and continue to experiment with a variety of reform measures, to respond positively to the current paradigm shift towards market-oriented economies and democratic society. An examination of the various reform .strategies undertaken worldwide provides a valuable "menu of options" for governments in a transitional reform stage or in the process of reconsidering the role of different sectors in agriculture.

REFERENCES

Axinn, G.H., 1988. *Guide on alternative extension approaches. Agricultural Education and Extension Service*, Human Resources Institutions and Agrarian Reform Division. Rome, FAO, M/S8971E/1/12.88/1000.

Morss, E.R. and D.D. Gow, 1985. Implementing rural development projects: lessons from AID and World Bank experiences. Boulder, CO (USA), Westview Press

FAO, 1981. The peasants' charter. The Declaration of Principles and Programme of Action of the World Conference on Agrarian Reform and Rural Development. Rome, FAO

4

Demand-driven Agricultural Extension

In recent years, many developing countries have reaffirmed the essential role that agricultural extension can play in agricultural development. This renewed interest in extension is linked to the rediscovery of the role that agriculture needs to play in reducing persistent rural poverty. Yet negative experiences with extension in the past have sparked considerable debate worldwide about the best way to provide and finance agricultural extension. What are the roles of the public sector, the private sector, and the third sector—nongovernmental organisations (NGOs) and farmer-based organisations—in providing and financing extension? How will the agricultural sector meet new challenges, such as helping smallholders access global markets and meet their standards? How can farmers cope with environmental degradation and climate change and respond to health challenges such as livestock pandemics? How can extension address the needs of women farmers and disadvantaged groups? What are the best uses of new information and communication technologies? All these questions are highly relevant for agricultural development in India.

In addition, agricultural extension faces the challenge of establishing a well-managed, effective, and accountable system that meets the needs of hundreds of thousands of farmers engaged in diverse and complex farming systems; the associated problems of monitoring and evaluating extension services and assessing their impacts; the dependence of extension on the performance of the agricultural research system and its feedback linkages; and the inherent problems of ensuring political commitment and fiscal accountability for agricultural extension.

The way in which agricultural extension has been organised and provided to meet these challenges has changed over time, with remarkably similar trends across the globe. These changes have been linked to general trends in development thinking and practice. A strong belief in the role of the state—as the major actor of development—characterised the economic policies of many developing nations after they reached their independence. The establishment of public sector extension services fitted well into this paradigm. The Training and Visit (T&V) system, promoted by the World Bank in more than 50 countries, became a major model for providing and managing extension. The disenchantment with the role of the state in development—reflected in the structural adjustment programs of the 1980s and 1990s—led to a downsising of agricultural extension in many countries.

The T&V system was finally abandoned in the late 1990s. The question remains whether the major reason was the lack of fiscal sustainability, the inadequacy of the model for many situations in which it was promoted, or its inconsistency with the growing emphasis on the role of the state as a facilitator rather than a provider of public services. In the post-T&V era, thinking and practice about extension has moved toward pluralistic modes of providing and financing extension. Major reform trends around the world include decentralisation, contracting, privatisation, cost recovery, and the involvement of NGOs and farmer-based organisations. Emphasis is now placed on making advisory services demand-driven. The emphasis on demand-driven extension has to be seen in the context of changing domestic and external environments for agriculture, which change the information needs of farmers. In addition to information on new technologies, advice on marketing, product quality, and environmental challenges has become increasingly important.

The concept of demand-driven services is also linked to a paradigm shift in public sector reform toward responsive governance. This paradigm emphasizes the need to make service provision accountable to users and to promote transparency and empowerment as essential conditions for increasing the effectiveness of public service provision. Thus, the focus on demand-driven service provision is not limited to agricultural extension: A similar emphasis can be observed in many other sectors. Efforts to make health care, education, and community water supplies demand-driven are also under way.

These general trends in agricultural extension can be observed in India. The T&V system played an important role in the Green Revolution. However, it was not well suited for the diverse farming system of rainfed areas and proved incapable of meeting the challenges of the post–Green Revolution period, including improving the sustainability of India's farming systems, promoting agricultural diversification, and integrating farmers into dynamic markets. As in other countries, many new approaches to providing and financing extension have been tried in India's post-T&V era.

In 2000, the Extension Division of the Department of Agriculture and Cooperation of the Ministry of Agriculture developed a draft of its Policy Framework for Agricultural Extension, which aimed for a major reform and reorientation of India's entire agricultural extension system. In line with the shifting international paradigm, the framework emphasizes pluralistic agricultural extension and the "promotion of demand-driven and farmer-accountable extension". It served as a basis for consultations with state governments and private sector organisations such as the Federation of Indian Chambers of Commerce and Industry. The document serves as the basis for India's extension policy and guides both internally and externally funded programs through which the government supports the states in their extension reform efforts.

The objective of this chapter is to discuss various strategies for making agricultural extension demand-driven, using India's Policy Framework for Agricultural Extension as an example. For this purpose, we develop a system for classifying options for providing and financing agricultural extension that involves the public, private, and third sectors. Each sector is subject to a unique set of challenges and failures. As our discussion shows, market failures, state failures, and community failures are widespread in agricultural extension. Various approaches address these failures, but they too are subject to challenges. In other words, there is no "free lunch" in reforming agricultural extension. We review India's Policy Framework for Agricultural Extension with regard to the strategies it identifies to make extension demand-driven, and we examine the provisions the framework makes to overcome market, state, and community failures. We also consider evidence from the 2003 Situation Assessment Survey of Farmers conducted by India's National Sample Survey Organisation and other data to identify the extent to which the framework actually addresses the problems revealed by the survey. Because other public services are also attempting to become more demand-driven, the relevance of our findings goes beyond agricultural extension services.

CONCEPT OF DEMAND-DRIVEN SERVICES

The term *demand-driven* refers to the economic concepts of supply and demand. In economic theory, *demand* refers to the amount of good or service that a consumer is willing and able to buy at a given price. Agricultural extension is characterised by various market failures that affect both the supply side and the demand side of advisory services. In view of these market failures, the public sector and the third sector have traditionally played a major role in financing and providing extension services. In this chapter, the *third sector* is defined as comprising civil society organisations, including farmer-based organisations and NGOs. Among the NGOs, non-profit providers of extension services are of particular interest. In the absence of the market mechanism, public and third sector extension providers face considerable challenges to ensuring that the services they supply meet the needs and the priorities of their clients. The term *supply driven* captures the criticism that this challenge has not been met. The T&V system, for example, is typically described as a supply-driven or top-down model.

The concept of demand-driven extension emphasizes the need to provide services that meet the needs and priorities of farmers, even if the market mechanism—Adam

Smith's famous "invisible hand"—fails to make sure that extension services are supplied in the quantity and quality expected by farmers. The term *farmer-driven* or *farmer-led extension* might better address the goal of making these services meet the needs and priorities of farmers, even if they are not able to exercise *demand,* as the term is defined in economics.

Because agricultural extension is a major area of donor funding and demand-driven public services are high on the international agenda, it is not surprising that the Neuchâtel Group—an international donor coordination forum on agricultural extension—has published guidelines on demanddriven agricultural advisory services. The group's publication recommends strategies to facilitate the emergence of a market for extension services and strategies to increase the voice of farmers when the public sector is financing and/or providing extension services. The publication also highlights the need to strengthen the capacity of farmers to articulate demand, as well as the capacity of service providers to respond to farmers' demands.

To establish demand-driven advisory services, it is useful to begin by identifying the extent to which market failures or other obstacles prevent the emergence of private sector extension services, which use the market mechanism to make services demand-driven. If an advisory service had no market failures and the market led to a socially desirable outcome, creating an environment in which the private or third sector could provide these services would be sufficient to make extension demand-driven. However, market failures in agricultural services are widespread. Hence, the question arises as to how demand-driven, or farmer-driven, advisory services can be established when the market mechanism fails to make these services demand-driven.

To identify strategies for making agricultural extension demand-driven, it is useful to consider the range of institutional options by which these services can be provided and financed, taking into account that organisations of the public, private, and third sectors can collaborate in various combinations.

EXTENSION SITUATION IN INDIA

Before discussing the provisions of India's Policy Framework for Agricultural Extension, a brief overview of extension in India would be useful. Extension played an important role in promoting Green Revolution technologies, and the T&V system proved effective in the areas of India affected by the Green Revolution. However, it was less effective in the rainfed areas. There is a general perception that, after T&V was phased out in the 1990s, the existing extension system deteriorated, even though a variety of new approaches to provide and finance extension emerged. These approaches included decentralisation; contracting; group extension; provision of extension by para-extension workers, producer cooperatives, or farmer-based organisations; the establishment of agro-clinics by private entrepreneurs with initial government support; public-private partnerships in financing and providing extension; and the establishment of Internet-based extension though village kiosks (e-Choupals) set up by the private sector. The Agricultural Technology Management Agency developed a model that embodies several of these reform elements. In view of low agricultural growth rates and a vigorous political debate about agrarian distress, the political attention to agricultural extension also was renewed. As one of the major "thrust areas" in agriculture, India's 10th Five-Year Plan (2002–2007) emphasizes the need for "revamping and modernising the extension systems and encouraging the private sector to take up extension services". As indicated earlier, the national Policy Framework for Agricultural Extension was developed in 2000 to serve as the basis for the central government's support to the states. In India's federal system, the states have the major responsibility for agriculture, including agricultural extension.

Table 1. Percentage of farmer households reporting good quality information on modern agricultural technology provided by various sources

State or Union Territory	*Extension Worker*	*TV*	*Radio*	*Newspaper*	*Input Dealer*	*Other Progressive Farmers*
Andhra Pradesh	45.8	60.4	44.6	53.8	57.0	54.9
Assam	30.5	40.1	36.4	47.8	28.4	24.8
Bihar	10.4	45.4	53.0	47.8	59.2	59.4
Chhattisgarh	38.6	63.5	82.8	52.8	0.0	52.5
Gujarat	60.3	73.7	59.2	56.4	73.4	66.4
Haryana	59.8	59.5	65.4	60.8	54.1	60.4
Jammu and Kashmir	18.0	44.6	36.6	49.8	8.4	52.1
Jharkhand	100.0	31.0	46.9	36.2	20.6	14.5
Karnataka	48.0	64.7	54.6	54.3	53.1	53.6
Kerala	51.0	60.6	62.1	62.3	70.6	57.7
Madhya Pradesh	42.1	66.2	62.5	57.4	51.9	34.6
Maharashtra	48.3	57.8	54.2	53.1	45.3	56.1
Orissa	57.3	48.4	46.2	43.8	26.0	24.4
Punjab	51.6	67.8	63.4	70.5	44.8	85.5
Rajasthan	70.8	63.6	58.7	71.0	42	42.5
Tamil Nadu	78.7	69.2	68.8	64	70.7	68.5
Utter Pradesh	35.9	55.2	63.6	54.4	52.9	58.6
West Bengal	45.0	52.1	46.0	56.8	36.7	41.2
India	51.1	59.0	55.5	55.9	50.5	52.8

The 59th round of the Situation Assessment Survey of Farmers carried out by the National Sample Survey Organisation (NSSO) in 2003 provides important information on the prevailing situation when the framework and the 10th plan were developed. Data for the survey, which is representative at the state level, were collected from 51,770 households in 6,638 villages.

Figure 1 displays the information sources for modern technology used by farmers in India. Farmers were asked to identify which, if any, of the sources they had accessed during the past 365 days to obtain information on modern agricultural technology. Nearly 60 percent of the farmers had not accessed any sources. When farmers did use sources, the input dealer was the second-most-used source after other progressive farmers. Output buyers, food processors, and credit agencies also fall in this category, if the respective companies are private sector entities.

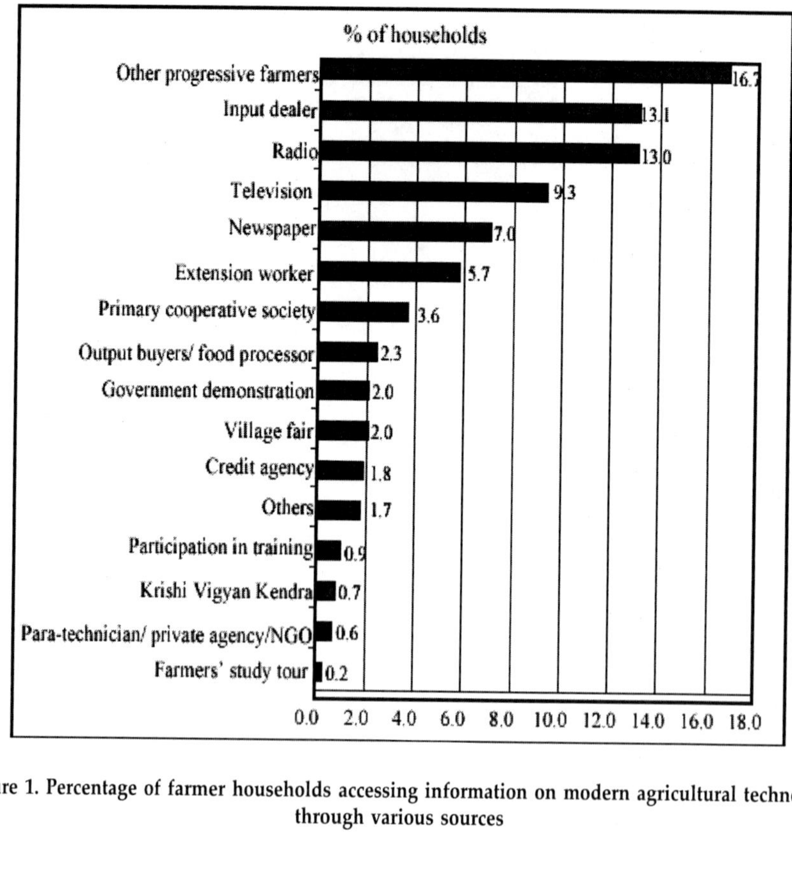

Figure 1. Percentage of farmer households accessing information on modern agricultural technology through various sources

The surveyed farmers were also asked to rank the quality of information received as good, satisfactory, or poor. About half of the farmers ranked the quality of information received from most of the information sources as good. There are, however, considerable differences across states. The percentage of farm households who rated the quality of the information provided as poor was actually rather small for all types of information sources reported. When asked to make suggestions for improvement, about one-third of the farmers mentioned improvement in the quality and reliability of the information provided. For all information sources, this suggestion was more frequently mentioned than any other, including improved timeliness of information.

MAKING EXTENSION DEMAND-DRIVEN

This section discusses the challenges of making various types of agricultural extension demand-driven and outlines strategies to meet these challenges. The provisions made in the Policy Framework for Agricultural Extension are then examined from this perspective. To identify the reasons for market failures, this section mainly draws on economic theory, while the broader agricultural extension literature is used as a basis for discussing the challenges and failures of the public and third sectors. The analysis of demand-driven services is best approached from a multidisciplinary perspective. In particular, management sciences and the innovation systems literature, which focus on organisational learning, knowledge management, partnerships, and institutional change, can provide important analytical insights for approaches to make extension demand-driven.

Market-Based Extension

Reasons for market failures

Market failure can be defined as the inability of a market production system to provide a good or service either at all or at a level that is optimal from the society's perspective. Imperfections in the market mechanism can be caused by the nature of the goods to be provided or by positive and negative externalities. Market failures can affect both the supply side and the demand side of service provision.

One reason for market failures in extension is that some types of information are public goods. Knowledge that is not farm specific, such as information about prices, is considered a *public good*, especially if it is distributed using a nonexcludable technology, such as the radio. Information provided through the Internet can be made excludable more easily—for example, by requiring that farmers become members of a group to access password-protected Web sites. This kind of service represents a *club good*. Farm-specific advice has the characteristics of a *private good*. With a private or club good, the nature of the good does not cause market failures.

Market failures also occur if an extension service has the characteristics of a *merit good*—a good that consumers undervalue because of imperfect information. Lacking sufficient information about the value of extension, farmers are likely to demand less of it than is in their best interests. Extension services may also have the character of merit goods because of time-horizon problems. Poor farmers with a high time rate of discount may undervalue the benefits of extension if those benefits are only realised later. This problem is aggravated when the benefits of extension are perceived as uncertain and farmers are risk averse. While the nature of public goods leads to market failures on the supply side, the character of extension as a merit good leads to market failures on the demand side.

Externalities are another reason for market failures. Extension is associated with positive externalities if the benefits of extension accrue partly to the society as a whole. An example is the use of extension for reaching national goals such as food security. Extension services also have positive externalities when they contribute to reducing the negative environmental effects of agricultural production. This can also be seen as the use of extension to reduce the negative externalities of agricultural production. Educating farmers about the negative environmental effects of production and promoting technologies that help avoid those effects will not necessarily lead to a full internalisation of negative externalities.

The impact of education depends on the farmers' environmental preferences, the differences between public and private costs, and the benefits of agricultural production systems that differ in their environmental impacts. The positive and negative externalities described here prevent market mechanisms alone from leading to a socially optimal provision of extension services because the private demand for them is insufficient. The characteristics of smallholder agriculture in developing countries may also lead to market failures. Because provision of extension is subject to economies of scale, providing extension services may be profitable for private entrepreneurs only if they can reach a sufficiently large number of farmers.

Agricultural producers are more spatially dispersed than urban populations, and they often have a more limited access to transportation infrastructure. Moreover, they are often not organised in groups. As a result, the transaction costs of providing extension to smallholders in less-developed areas are typically high, and private sector organisations may not find it profitable to provide those services. In early phases of agricultural development, the same factors (spatial dispersion, lack of coordination, high transaction costs) also give rise to market failures in other agrarian institutions, such as agricultural credit and marketing. These market failures may lead to associated market failures in extension. For example, farmers may not have access to credit to pay for extension services. These problems are usually reduced in the course of economic development, thus creating scope for market-based extension services to arise. In contrast, the market failures resulting from externalities and public goods are not dependent on the stage of economic development.

Market failures are not the only justification for the public sector to intervene in agricultural extension. The state may choose to finance agricultural extension as a component of a poverty alleviation strategy. Institutions such as contract farming that primarily address market failures in agriculture other than those specific to extension may also provide a solution for extension. Extension services can be embedded in contracts that integrate farmers into value chains.

Provisions in the Policy Framework

The framework acknowledges the relevance of market failures in agricultural extension by pointing out that "pure public goods, economically backward regions, small, marginal farmers and landless labourers will not attract the for-profit private sector. Public Extension will therefore continue to play a central role in technology dissemination". At the same time, the framework emphasizes the need for "withdrawal of the public sector from areas where agro-services can be effectively and competitively provided by the private sector" to make sure that public sector provision of extension does not crowd out private extension providers. However, given that less than 6 percent of farmers used public sector extension agents as sources of information, it is doubtful that crowding out played a role in the low prevalence of private sector extension. Unlike in other policy areas, such as agricultural marketing, India has no government regulations preventing private extension providers from operating. The framework mentions artificial insemination services, soil testing, fertiliser advice, farm improvement plans, and breeding plans as examples of "private goods" for which competitive markets exist. Moreover, the framework recommends using contract farming "wherever feasible," particularly in the area of high-value and export-oriented agriculture.

Public Sector Extension

State failures

State failures in agricultural extension occur because of problems related to information, incentives, capacity, political interests, and bureaucratic procedures and attitudes. Although these problems are not specific to agricultural extension, the complexity of smallholder agriculture aggravates some of the typical public sector problems. From the perspective of the new institutional economics, agricultural extension is transaction intensive. Moreover, moving from the delivery of standardised messages to demand-driven advice makes extension more discretionary and specific. Public services that are both transaction intensive and discretionary are particularly difficult to provide. Two types of *information problems* lead to state failures in agricultural extension. First, in the absence of a market mechanism, public sector extension providers have trouble determining the types of knowledge and advice farmers actually need. Making this determination is at the heart of making public sector extension demand-driven. An information problem that extension managers often face is the inability to determine what extension agents actually do in the field and to supervise them. Considerable information asymmetries exist between extension agents and their managers because of the spatially dispersed nature of agriculture and because an agricultural outcome, such as crop yields, are influenced by many factors, other than extension.

Incentive problems also lead to state failures. Public sector agencies can use various instruments to create incentives, such as merit-based promotion. However, when public budgets do not keep pace with inflation and general improvements in standard of living, as is often the case, the possibilities to use a differentiated system of promotion associated with salary incentives deteriorate. In addition, extension agents often have a lower social status than many other public sector employees and a lower rank in the civil service system, which affect their morale. Another factor affecting agents' morale is that they do not have the operational funds they need to get to the field and work effectively. The incentive problems inherent in public sector extension agencies are closely linked with *capacity problems*. Public sector agencies often lack the incentives to invest in the capacity of their extension staff so as to keep their knowledge up-to-date.

Another failure inherent in public sector extension is *political interest capture*. Because largescale farmers often have more political influence than smallholders, politicians have an incentive to induce the public administration to serve large-scale farmers and to concentrate on extension issues that are more relevant to them. Often the only government agents able to interact with a considerable part of the rural population, extension agents may be misused for political purposes, such as campaigning for the ruling parties in elections.

Public sector extension agents are also often burdened with other activities that are outside the mandate of agricultural extension. For example, extension agents have frequently been involved in organising the supply of subsidised inputs and implementing credit schemes. They may also be asked to get involved in implementing public health programs and other government schemes unrelated to agriculture. Corruption is not a typical problem in extension because knowledge services do not offer much scope for corruption. However, the more extension agents are involved in the distribution of inputs and credit or in the enforcement of laws, the greater is the opportunity for corruption. *Bureaucratic procedures* make it difficult for extension agents to respond flexibly to local demands, especially in highly centralised systems. *Bureaucratic culture* is a typical obstacle to the reform of public sector agencies.

Encouraging processes of institutional learning and change is a major challenge in public sector agencies. Likewise, *bureaucratic structures* often discourage the coordination of agricultural extension with other departments. Even links to the agricultural research system are often weak in spite of their obvious importance. Farmers may also suffer from *attitudinal problems* that are often widespread in traditional public sector agencies. Rather than treating farmers as clients, customers, or citizens, extension agents too often consider them state subjects.

Typically, farmers who are poor, female, or part of some socially excluded group are particularly affected by attitudinal problems. A culture of mistrust developed from attitudinal problems can also be an obstacle to building partnerships with the private sector and civil society organisations. *Financial sustainability* is another problem of public sector extension, especially if cost recovery is not pursued. After donor-funded programs end, extension agencies are often left with an increased number of agents. Because staff numbers are difficult to reduce in public sector agencies, budget reductions limit the resources available to extension agents to do their jobs effectively, such as transport facilities and training. These problems lead to negative feedback. Without sufficient resources, public sector extension services become ineffective, and their image of being ineffective leads finance ministries to deny requests for more resources.

Last but not least, public sector extension providers can cause the *crowding out* of private and third sector extension providers. This problem occurs when government agencies provide extension free of charge even though no market failure has occurred. Governments may also establish regulatory or bureaucratic barriers that prevent private sector extension agencies from emerging. This type of problem, however, seems more prevalent in the areas of government-supported input supply and marketing than in the actual provision of extension services. Hence, this type of government failure is more likely to affect extension services provided by input dealers and output buyers or processors than those provided by private extension agencies.

Strategy to Address State Failures

One important strategy to address state failures in agricultural extension is to involve NGOs, farmerbased organisations, and private sector agencies in the management and execution of extension services. The present section deals with strategies that can be used within the public sector to overcome state failures and make extension more demand-driven. These strategies fall into four areas: institutional design, funding mechanisms, management approaches, and extension methods.

Institutional design

Decentralisation is an important strategy to make public agencies more responsive to local needs. It can take two forms: *devolution,* or making public agencies accountable to locally elected governments, and *deconcentration,* or transferring authority to offices at lower levels of government but retaining accountability within the line agency. Decentralisation involves its own challenges. Problems of political interest capture and incentives to burden extension agents with other tasks may increase after decentralisation. If funding responsibilities are transferred to local governments, extension may no longer be a priority, especially when basic needs, such as water, education, and health, are not being met.

The institutional design of public sector extension agencies can help overcome some of the state failures discussed earlier. *Increased autonomy* for extension agencies can be an important approach to reducing political interest capture and limiting opportunities to burden extension agents with tasks outside their mandate. The challenge of increased autonomy is "delegatee drift": A more autonomous agency may not necessarily pursue the public goals that policymakers intended in setting up the agency.

Contracting is an important strategy to address state failures by institutional design. If contracting is done through competitive bidding, the competition mechanism can be used to address some incentive problems. However, contracting involves considerable challenges, because the public sector needs to manage the contracting process, which involves all the problems inherent in procurement, including corruption. Whether the public sector is better able to overcome the state failures inherent in managing its own extension agents than the state failures inherent in procurement is an empirical question.

Funding mechanisms

The way funding is provided to public sector agencies can affect incentive problems. A mechanism widely used in agricultural research, but less so in agricultural extension, is the competitive grant. An important strategy to improve both financial sustainability and demand orientation is cost recovery—for example, by charging a fee for participation in extension activities. It may be politically difficult to move to a fee-for-extension scheme. The willingness or ability to pay, especially among poor farmers, may be constrained by market failures. While having to pay a fee increases farmers' incentive to hold extension providers accountable, it is not necessarily a mechanism that makes extension providers accountable to the farmers. In the absence of a market mechanism, it is necessary to establish institutional channels by which farmers' demands are translated into management decisions. As long as farmer-toagent ratios are more than 1,000:1, establishing such mechanisms necessarily involves farmer-based organisations rather than individual farmers.

Management approaches

The public sector can use a range of managerial approaches to address the problem of weak incentives. As indicated earlier, merit-based recruitment and promotion is one of the most important strategies in this regard. Other instruments include performance contracts and other "managing for results" approaches, seeking feedback through client satisfaction surveys, establishing professional standards, and other efforts to promote a "mission-oriented" service. The new public management approach aims at introducing a range of private sector management techniques to public administration. Because public sector extension is typically part of the general public administration, the opportunities to use such instruments are often constrained by formal civil service rules and by an informal bureaucratic culture. Changing such formal and informal rules for extension in isolation from the rest of the bureaucracy is likely to be difficult. The creation of semiautonomous agencies is one strategy to increase the scope for applying management approaches that aim to resolve incentive problems. All these strategies can be considered supply-side approaches.

Extension methods

Extension methods differ widely with regard to the scope they create for allowing farmers to articulate demand. On one end of the spectrum are transfer-of-technology methods that aim at disseminating new technologies developed in research stations, such as lectures and instructions, demonstration plots, and information dissemination via radio. Such methods—which were used, for example, in the T&V system—leave limited room for the articulation of farmers' demands. On the other end of the spectrum are participatory extension methods, including participatory technology development and the Farmers' Field School approach, which create more space to tailor extension to the demands of farmers. One challenge often faced by public sector extension agents is that the use of participatory extension methods requires the development of specific skills, such as group facilitation. Without investing in training for extension agents to develop such skills, it is difficult for the public sector—as for any other extension provider—to use the potential inherent in participatory methods.

Provisions in the policy framework

The framework does not include a systematic description of the state failures the framework seeks to address. However, it mentions a range of problems. The challenge of financial sustainability and limited operational funds is explicitly mentioned: "States have barely been able to pay the salaries of extension personnel. Less than 10 percent of the budget is available for operational expenses, which has practically immobilised the service with scarcely any technology dissemination in the field". In reviewing earlier extension approaches, the framework criticises the "top-down nature of the Training and Visit System" and observes that "farmer driven and farmer accountable feedback systems were not adequately developed". The framework also acknowledges capacity problems. On the problem of crowding out, the framework states, "If the public sector continues to subsidise the services, this will prevent a 'level playing field' to the private sector, which will ultimately get crowded out". The framework also stipulates "a re-examination of existing Rules, Regulations & Acts to abolish provisions, which constrain private investment in delivery of agro-services".

The framework places strong emphasis on decentralisation, in the form of both deconcentration and devolution. The Agricultural Technology Management Agency (ATMA) model is proposed as the key concept for decentralising decision making to the local level. However, the framework itself does not provide further specifications of the ATMA model, which was originally introduced under an agricultural technology project funded by the World Bank. The explanations presented here are based on project documents and other publications.

The governing board comprises the heads of various line departments and research units as well as stakeholders—including a cross section of farmers, women, and disadvantaged groups—and private sector firms within the district. The Farmer Advisory Block Technology Team comprises personnel with extension functions from various departments. The Farmer Advisory Block Technology Committee plays a key role in ensuring bottom-up planning. It is composed of the head of the Farmer Interest Groups, which are typically organised around specific crops. Farm Information and Advisory Centres provide a single-window delivery mechanism for extension. Owing to this setup, the ATMA promotes both coordination among government agencies and coordination among the public, private, and third sectors.

In its original design, the ATMA may be considered a model of deconcentration rather than devolution because, at least in its original implementation, elected local government leaders were not part of its governance structure. However, the framework acknowledges that local governments—specifically, India's three-tier Panchayati Raj Institutions—have started to play a role in extension. As stated in the framework, "Some states have also delegated suitable administrative and financial powers to the institutions. In these states, the extension personnel are placed under the administrative control of the local governments (*panchayats*), whereas for technical guidance they remain under the control of their respective technical line departments".

Accordingly, the framework encourages the development of linkages between the ATMA units at village, block, and district levels and the evolving Panchayati Raj Institutions. Moreover, as part of the ongoing decentralisation process, the Ministry of

Agriculture expects to discontinue 27 centrally funded programs and reallocate those funds directly to the ATMA districts.

The framework suggests the strategy of increased autonomy discussed earlier. The ATMA model ensures a considerable degree of independence from the general public administration: The ATMA is a registered society, which has much flexibility, for example, to enter into partnership with private sector entities. The framework also promotes the principle of increased autonomy for the State Agricultural Management Extension Training Institutes by making such autonomy a precondition of support from the central level.

The framework envisages "contracting out of extension services to private-sector, community-based organisations or NGOs in selected geographical areas (e.g., a village, cluster of villages, Block)". The framework acknowledges the need for a "transparent, laid out procedure" and for a "strict monitoring and evaluation process". Additionally, the framework points out that extension support services, such as security, mobility, and computer and secretarial services should be contracted "wherever possible". In particular, participatory planning should be contracted to NGOs; staff training, to universities or institutes; and monitoring, to farmer-based organisations, the Indian Institute of Management, and other institutions. Moreover, the framework states that, in contracting arrangements among governments, extension services, and farmers, "the farmers could play the role of beneficiaries, provider or co-financier of extension services". Related to contracting and the increased roles of the private sector, the media, and information technology, the framework stipulates that the public extension service should be made "leaner and professional. It is envisaged that the approximately 100,000 public extension functionaries will be gradually reduced to be supported by the other two arms of services providers".

The framework suggests both the establishment of competitive grants and cost recovery mechanisms. It envisions a Competitive Agriculture Extension Grant Fund (CAEGF) and suggests that public extension agencies compete with private extension agencies for operational funds under such a grant. An independent impact evaluation is suggested to assess performance as a basis for subsequent eligibility to compete for funds. The Competitive Agriculture Research Grant Fund set up in the Indian Council of Agricultural Research and several states is mentioned as a model for the proposed CAEGF. The framework encourages cost recovery: "Wherever farmers have the capacity to pay for public services, which are in the nature of private goods, realistic cost of such services should be recovered. However, provision is made for targeted subsidies to protect the vulnerable class of users". Likewise, the framework states that the private extension providers should charge for extension services or, in case of contract farming, recover the costs out of their profit margins. Innovations in financial institutions, such as revolving funds, and linking farmers with credit institutions are also encouraged by the framework.

The framework addresses the problem of financial sustainability by suggesting cost-cutting mechanisms for extension services: "Cost effectiveness may be improved by relying on fewer but better qualified (graduate or post-graduate) field advisers who interact directly with researchers for subjectmatter advice and then multiply their impact in the field by working with farmer groups rather than individual contact farmers".

The framework creates the conditions for using incentive-oriented management approaches by increasing the autonomy of the extension agencies through the ATMA model, as indicated earlier. It also encourages the states to develop human resource development (HRD) policies and action plans by making them a precondition for central funding. According to the framework, HRD policies "would also build in an effective system of rewards and incentives for public extension functionaries". However, improving merit-based recruitment and promotion and adjusting the pay scale of public sector extension agents to create incentives are not specified in the framework. Neither does the framework refer to the option of linking HRD reform efforts in extension to those in general public administrations. The framework does, however, emphasize capacity strengthening. It stipulates increasing the professional qualification of extension staff by suggesting a bachelor of science degree in agriculture as the minimum educational requirement for farm advisors. "DOA's extension field staff would be restructured and upgraded to create a professional cadre of Farm Advisors. In the process, the village extension worker (VEW) cadre would be incrementally phased out through reassignment and normal attrition". To improve the skills of the extension agents who will remain in the public sector, the framework advocates a "massive campaign" comprising a skill gap analysis and compulsory training in the form of foundation courses and a professional course.

The framework strongly recommends participatory extension approaches by encouraging participation in "working out the system description, problem diagnosis, search for appropriate technology, designing the process of implementation, monitoring and evaluation, and feedback". Likewise, the framework stresses the role of indigenous knowledge and partnership, highlighting that "the extension agent is no longer seen as the expert who has all the useful information and technical solutions; the indigenous technical knowledge of farmers and their ingenuity, individually and collectively, are recognised as a major source; and solutions to local problems are to be developed in partnership between the extension agent and farmers". The framework also suggests the use of group extension as a mechanism to make extension demand-driven. Farmer interest groups should "first generate a demand for information, technology and management techniques". The need for capacity building to enable extension agents to use such methods is fully recognised in the framework: "Extension workers therefore need new skills of negotiation, conflict resolution and mobilising and nurturing community organisations".

Third Sector Extension

It is useful to distinguish two types of third sector organisations that can be involved in agricultural extension: NGOs, especially non-profit service providers, and farmer-based organisations (FBOs), which are membership organisations formed by farmers. Whereas NGOs are accountable to their funding agencies, FBOs are accountable to their members.

Potentials of NGO Extension

One of the major advantages of NGOs is their flexibility. Because they are not part of the public administration, they are not constrained by civil service rules and are usually less confined by bureaucratic procedures and cultures. Therefore, they can use a range of managerial approaches to create incentives for their staff, such as merit-based and competitive payment, which are often not available to the public sector. Depending on its type, an NGO may be in a better position to attract staff members who are intrinsically motivated and dedicated to its cause. Moreover, an NGO is not usually subject to the problem of political interest capture and is typically less likely to be under pressure to assume responsibilities outside the mandate of agricultural extension. NGOs have often played a pioneering role in introducing group-based and participatory extension methods. In contracting systems, NGOs may compete among themselves or with for-profit private sector providers, which may create additional incentives for improved service provision.

NGO failures

When NGOs work in the absence of a market mechanism, they are, in principle, subject to the same types of information problems as the public sector. They are usually small compared with public sector agencies. Although their size contributes to their flexibility, it also reduces their outreach. In fact, the percentage of farmers who are reached by NGO extension is often small. NGOs constituted an information source for less than 1 percent of Indian farmers. Some NGOs have grown to a considerable size, such as BRAC in Bangladesh.

However, the larger an NGO becomes, the more likely it will become subject to the same problems of bureaucracy faced by public sector agencies. Moreover, NGOs are subject to accountability problems. In principle, public sector agencies can be made accountable to farmers through political channels. NGOs are only accountable to their funding agencies. They often face problems of financial sustainability and have to cope with considerable fluctuations in funding. Moreover, NGOs are not immune to problems of mismanagement and misuse of funding. Neither are they immune to attitudinal problems.

While NGO staff would not treat farmers as "state subjects," they might become patronising toward those they serve. Additionally, they can have incentives to combine extension with other goals, including promoting the worldview or religion with which the NGO is associated. Frequently, NGOs do not have their own extension staff but use public sector extension agents. If this fact is neglected, their contribution to increasing the supply of extension services is easily overestimated.

Strategies to overcome NGO failures

To overcome the challenges they face, NGOs can to some extent use the same strategies discussed earlier for public sector agencies. For example, NGOs can use participatory extension methods and involve FBOs in extension management to solve information problems. Likewise, they can use the same types of managerial approaches mentioned earlier for the public sector, such as making use of client satisfaction surveys and focusing on results-based management. NGOs also can use cost-recovery methods, such as fee-based extension, to improve financial sustainability and increase farmers' incentives to demand accountability.

Provisions in the policy framework

The framework foresees an important role for NGOs. It highlights their "ability to mobilise communities into Farmers Organisations/Farmer Interest Groups/Watershed Associations/Market Associations" as a major strength. According to the framework, NGOs can either complement public sector extension by focusing on community mobilisation or substitute for public extension through contracting approaches. The central government is also accorded a role in partnering with NGOs. As the framework states, of the then 261 Farmer Information and Advisory Centres in the country, 86 were operated by NGOs.

The framework also proposes the use of public funding for capacity building of NGOs: "A systematic training, capacity building and technical backstopping mechanism, supported through public funds is to be developed for NGOs involved in providing extension services". Apart from acknowledging the need to provide training, however, the framework does not refer to the challenges involved in NGO extension or to strategies to deal with those challenges.

Potentials of FBO Extension

FBOs have important advantages, especially for smallholder agriculture. As mentioned earlier, the farmer-to-agent ratio in most developing countries is more than 1,000:1. Hence, it is difficult for farmers to exercise demand and hold service providers accountable without some form of organisation. FBOs can play an important role in aggregating farmers' demands for extension and in representing farmers in participatory models of extension management. For example, FBOs can represent farmers in developing plans for extension. They can also represent farmers on procurement boards charged with contracting extension providers and can participate in multistakeholder management boards. Moreover, FBOs can help reduce the transaction costs of providing extension, as in group-based extension approaches. FBOs can partner with public sector agencies as well as NGO service providers to make extension more demand-driven. An important advantage of FBOs is the fact that they are, in principle, directly accountable to the farmers who are their members.

FBO failures

A major challenge for FBOs is the classical problem of collective action. If the benefits of FBO action are "nonexcludable," farmers have limited incentives to incur the transaction costs of participating in an FBO. The incentives to join local farmer groups for the purpose of group-based extension may be rather high, because the participants expect to benefit directly from their participation. However, to participate in extension planning and management beyond the district level, farmers need to become organised at a more aggregate level, which poses its own challenges. The literature on the role of group size and heterogeneity in collective action is extensive, and the relations continue to be debated. In any case, it is a common phenomenon that organisations formed for donor-funded projects collapse once project funding ends.

FBOs are not equally suited for all kinds of extension. They have particular comparative advantages to facilitate extension for activities that require collective action, such as many types of natural resource management. Where more farm-specific advice is required, their comparative advantage is lower.

A major challenge of FBOs is avoiding social exclusion and elite capture. FBOs are often dominated by middle-class and wealthy farmers. Poor farmers and socially marginalised groups typically play a limited role in the leadership of FBOs, even if they are members. Moreover, the representation of women in FBOs is often low, a problem linked to the sociocultural role of women in most societies as well as to the time constraints faced by women.

Importantly, FBOs are often confronted with capacity problems, particularly when they become service providers and hire extension agents. This requires skills in management, accounting, and supervision. If FBOs want to access high-value markets and export markets, capacity challenges are particularly pronounced. Depending on their size and activities, FBOs may choose to hire professional managers, a practice that compounds supervision problems. In addition, FBOs that hire professionals confront some of the same problems NGOs face. For example, they are not immune to mismanaging funds. Agricultural extension is not necessarily a priority for FBOs. Lobbying for state support in the form of subsidies is often a higher priority than helping their members become more competitive. If FBOs become large, they are also likely to become subject to political interest capture, because politicians have incentives to use them for mobilising votes. Likewise, leaders of large FBOs may have incentives to run for political office, which in turn may lead to conflicts of interest.

Strategies to overcome FBO failures

FBOs are rarely founded for the purpose of agricultural extension alone. Extension may be one activity in commodity-based FBOs, such as dairy cooperatives or vegetable grower associations. Integrating extension into FBOs that are founded for a broader set of goals is an important strategy to reduce the transaction costs of collective action.

Investment in the managerial capacity of FBOs through training can be an important strategy to overcome the management challenges they face. The dairy cooperatives in India, for example, have a high capacity to provide services to their members because of specific investments in their capacity by the National Dairy Development Board. Leadership training is also a promising strategy, because meeting the collective action problem requires vision and leadership. Networking, which allows FBOs to learn from each other, is another approach to overcoming FBO failures. One strategy to deal with the elite capture and social exclusion problem is the formation of specialised organisations, such as a group exclusively for women farmers. Institutional design can also help to address some of the problems faced by FBOs. For example, women and disadvantaged groups may be allocated seats in participatory planning and management boards for extension. In developing strategies to overcome FBO failures, it is important to keep in mind that donors and the public sector can play only limited roles. Ultimately, the formation of sustainable FBOs requires a process of social mobilisation in which the farmers themselves have to play the lead role. Donor-funded initiatives can only serve as catalysts. Importantly, the state needs to create the fundamental conditions that allow FBOs to thrive, such as a conducive legal environment and safeguards against political capture.

Provisions in the policy framework

The framework identifies FBOs as a major mechanism to make extension services demand-driven. It defines several types of FBOs, including self-help groups, farmer interest groups, and commodity associations. The framework emphasizes that FBOs can provide "an effective channel for both the dissemination of technology" to large numbers of small and marginal farmers, and feedback to research and extension and points out that they are especially important "for high value commodities and resource poor farmers". Based on the ATMA model, FBOs play a key role in this extension approach.

The question of how to promote the formation of FBOs is addressed in the framework in various ways. First, the framework points out that government services can help identify and strengthen existing associations or cooperatives of farmers. Likewise, as mentioned earlier, NGOs' support in the formation and mobilisation of FBOs is encouraged. The framework also foresees the use of public funds to support the formation of FBOs and their involvement in extension planning, implementation, and monitoring. Linking FBOs to local governments through existing institutions such as land management committees and federating FBOs at higher levels is also encouraged by the framework. The "internalisation" of extension services by FBOs is cited as an "ultimate aim". The framework also considers FBOs a major avenue of farmer integration with agribusiness and considers "support to farmers' organisations" as "perhaps the main single input that governments can provide" in this regard. Although the framework addresses the collective action problem inherent in forming FBOs, it does not outline strategies to encourage small and marginalised farmers or female farmers to join FBOs. Likewise, the problem of local elite capture in FBOs is not addressed.

RELEVANCE OF THE STRATEGIES

Before discussing the relevance of the strategies to India's agricultural extension policies, it may be useful to summarise them. Three types of extension were distinguished: private sector (market based), public sector, and third sector. In private sector extension systems, it is the market mechanism that fosters demand-driven services. If a market for extension does not exist because government interventions have "crowded out" private extension providers, reducing such crowding out and creating a favorable investment climate for private providers is one strategy to make extension more demand-driven. To what extent crowding out, rather than other types of market failures, prevents the emergence of a market for extension is an empirical question. Besides the "pure" type of market-based extension where private sector organisations provide extension services and farmers pay for them, other market-based approaches include agricultural advice provided with the sale of inputs or purchase of products and advice provided in a contract-farming relationship. How demand-driven these approaches are depends largely on the bargaining power of the farmers in these relations.

Public sector extension is often not responsive to farmers' demands owing to a range of government failures. Strategies to make public sector extension more responsive to farmers' demands include decentralisation of extension agencies, increased autonomy of extension agencies, contracting extension services and involving farmers in awarding the contracts, using funding mechanisms such as cost recovery to encourage farmers to express their demands, using management techniques such as new public management to emphasize responsiveness to clients, and using participatory extension methods. Extension services provided by NGOs and FBOs are not necessarily demand-driven either, because they are subject to various failures, too. The strategies to make NGObased extension more responsive to farmers' demands are similar to those that the public sector can use. Strategies to make FBO-based extension more demand-driven include strengthening management capacity and the internal accountability mechanisms of FBOs, with a specific focus on overcoming problems of elite capture and social exclusion. Because FBOs can play an important role in making other types of extension more demand-driven and accountable, strengthening the capacity of FBOs to articulate farmers' demands is an important crosscutting strategy.

Each available strategy to make agricultural extension more demand-driven has its own challenges. In view of these challenges, the choice of strategies depends on underlying assumptions and value judgments, or paradigms, regarding the roles that the public sector, the private sector, and civil society should play in agricultural development and, more broadly, in economic development. The Policy Framework for Agricultural Extension represents the paradigm of public sector reform that emerged in the "post–Washington Consensus" era. This paradigm acknowledges the role of the state in overcoming market failure, but it envisions a facilitating and enabling role for the state.

When privatisation is not possible, outsourcing is considered a major solution, and cost-recovery is a central element of the paradigm outlined in the framework. Other core elements are user participation, accountability, and demand-driven services. Subsidies are accepted but only if targeted to the poor and to marginalised groups. Reforms within the public sector receive comparatively little attention the paradigm, based on an implicit scepticism regarding the ability of the public sector to reinvent itself. The spirit and the language used in the framework are remarkably similar to international documents that reflect this paradigm. For example, the 2004 World Development Report on public service provision promotes similar strategies of contracting, cost recovery, client empowerment, and avoiding the "crowding out" of nongovernmental providers. Because the framework has not been translated into laws at the state level, it remains unknown, however, whether it would have been approved by elected political decision makers in this form, and which of the different elements would have been contested by different political parties. However, the framework can certainly be considered a bold approach to reforming extension services in India.

The framework was obviously inspired by the reform experience of the ATMA model. Studies evaluating the impact of the pilot operations have not focused on the question of whether extension has become more demand-driven, but according to various unpublished World Bank project documents, they have demonstrated encouragingly positive results. In the 28 districts in which the ATMA model was implemented under an agricultural technology project financed by the World Bank, the following increases in cropped areas were reported for the 1999–2003 period: horticulture, 12–16 percent; oil seeds, 3–11 percent; herbs and medicinal crops, 1–5 percent. Cereal areas declined from 55 percent to 47 percent, but importantly, yields increased 14 percent. Recorded farm income increased an average of 24 percent in project districts compared with 5 percent in other districts; this increase is statistically and surely economically significant if it can be sustained when the project is scaled up. However, collection of more broad-based geographic data and careful comparisons with remaining nonproject cases are needed to substantiate these results.

So far, the extent to which extension in India has changed on a broad scale since the framework was implemented remains unclear. India's 10th Five-Year Plan stipulates expanding the ATMA model to 252 districts and extending it to all of India's 600 rural districts during the 11th Plan. No nationwide representative survey, similar to the 2003 Situation Assessment Survey of Farmers is available to describe the current situation. According to extension experts, private sector participation in extension has increased because of growing exports and quality considerations, but the role of public-private partnerships has made little progress. Insufficient recruitment has reduced the number of public sector extension personnel, leaving especially remote areas poorly served. Technology dissemination has apparently remained the major goal of public sector extension, and evaluation systems still focus on indicators such as the number of trainings conducted or the area covered by modern varieties.

Results from a survey of farmer households conducted by the Institute for Social and Economic Change (ISEC) and the International Food Policy Research Institute (IFPRI) on decentralisation and public service provision in Karnataka provide some additional glimpses. Twentytwo percent of the 966 farmer households surveyed said they had at least one contact with a government extension worker during the past year, a significant increase over the average of 11.5 percent reported for Karnataka in the 2003 Situation Assessment Survey of Farmers. However, the most important problem reported by the respondents in the ISEC-IFPRI survey was that the "extension worker does not visit the village" (76 percent) or "does not visit the village regularly" (18 percent). As in the Situation Assessment Survey respondents to the ISEC-IFPRI survey were either very satisfied (51 percent) or somewhat satisfied (27 percent) with the knowledge provided by the extension agent.

Group extension had apparently not become widespread in the survey area, because only 15 percent of all extension interactions were group based. To collect information on accountability and demand orientation, respondents were asked whom, if anyone, they contact to report problems. Ninety-four percent of survey participants who answered this question reported that they had not contacted anyone. Most respondents who had contacted someone had turned to a local government official. These findings suggest that the task of establishing functioning channels for demand-driven extension remains largely unfinished.

The insights from the Situation Assessment Survey and the Karnataka survey raise the question of how well the Policy Framework for Agricultural Extension is focused on the major problems of public sector extension in India. The framework's emphasis on capacity strengthening and on increasing the professional qualification of extension agents suggests that the quality of the knowledge provided by the extension agents is a major problem. Market-oriented extension and farm diversification certainly require an upgrade in the knowledge of extension agents. However, the empirical evidence suggests that farmers presently perceive access to extension to be a more important limitation.

In this regard, the question arises as to what would be an appropriate level of access in the Indian context. On the one hand, a low coverage may be a normal feature in agricultural extension, because extension agents cannot interact with all farmers and technology diffusion relies on the spread of information among farmers. Given that the most important source for information about new technologies is "other progressive farmers" this mechanism works to a considerable extent. However, the low percentage of farmers who interact with government extension workers may point to the problem of "absenteeism," which is quite widespread in other areas of public service provision in India, such as health and education.

It may be useful to compare this extension contact percentage with data from statistically representative surveys from other countries. For example, the Service Delivery and Satisfaction Survey III conducted in Malawi found that an average of 46 percent of farmers interacted with extension workers . Uganda's National Service Delivery Survey conducted in 2004 found that 14 percent of farmer households had been visited by extension workers within 12 months before the survey compared with about 29 percent reported in Uganda's 2000 survey. Considering that both countries are at a lower level of economic development than India, the figure of 6 percent from the Indian survey does indicate a low outreach of public extension services. Hence, it is an important question for further research whether the strategy proposed in the framework—to reduce the number of public sector extension personnel in combination with promoting group-based extension and private sector providers—will be effective in improving farmers' access to extension services in India.

It is surprising that the framework does not address the challenges faced by FBOs regarding elite capture and exclusion, especially in view of the large role that the framework accords these organisations. In India, hierarchical local power relations are widespread. Members of scheduled castes and tribes face social and economic exclusion. Hence, special provisions to make sure that they are adequately represented in FBOs would have been justified in the framework to make sure that these groups' demands for extension are equally considered. The same concern applies to women. The framework places strong emphasis on mainstreaming women in agriculture and proposes a wide range of measures to increase women's access to services. However, the proposed measures mostly target the supply side of extension—for example, by promoting gender-sensitivity training for extension agents and increasing the number of female extension staff. Special provisions to strengthen women's ability to *demand* extension services are surprisingly absent in the framework. The design of the ATMA model envisages the representation of women and disadvantaged groups in the district-level governing board, but such provisions are not mentioned in the framework.

The framework pays comparatively little attention to the challenges inherent in reforming public sector extension agencies, in creating incentives for the staff who continue to be employed by the public sector, and in changing bureaucratic cultures and promoting institutional learning and change. Although the framework acknowledges the need for public sector involvement in serving the poor and reaching remote areas, important challenges to achieving this goal, such as creating incentives for qualified staff to serve in remote areas, are not addressed. The framework emphasizes reform elements that are in line with the post–Washington Consensus paradigm, such as contracting and competitive grants, but little empirical evidence exists to show the circumstances under which these approaches improve extension services.

Given the size and diversity of India, it is also surprising that the framework does not place more emphasis on the need to tailor solutions to specific circumstances. What works to make extension demand-driven in Andhra Pradesh, for instance, may not work in Bihar. The framework seems to assume that ATMA-type approaches are the best solutions under all circumstances, even though there is some evidence that the ATMA model did not work equally well in all districts where it was piloted. An institutional analysis to determine the reasons for the variable performance was not conducted, and the different states face different challenges to fit the model into their institutional structures.Other interesting reform approaches are available, such as the Kerala Horticultural Development Program, from which important lessons on extension reform could be derived. Evidence from public sector reform in general, and from extension reform in particular, suggests the need to move from "one size fits all" to "best fit" approaches. Moreover, as analysis of the Indian extension system has pointed out, a focus on experimentation and learning is essential to improving agricultural extension in India.

REFERENCES

Anderson, J.R., G. Feder, and S. Ganguly. 2006 (May). *The rise and fall of training and visit extension: An Asian mini-drama with an African epilogue.* Policy Research Working Paper 3928. Washington, D.C.: World Bank.

Bardhan, P., and D. Mookherjee, eds. 2006. *Decentralization and local governance in developing countries: A comparative perspective.* Cambridge, Mass.: MIT Press.

Rivera, W.M. 1996. Agricultural extension in transition worldwide: Structural, financial and managerial reform strategies. *Public Administration and Development* 16: 151–161.

Swanson, B.E. 2006. The changing role of agricultural extension in a global economy. *Journal of International Agricultural and Extension Education* 11(3): 5–17.

Tripp, R., M. Wijeratne, and V.H. Piyadasa. 2005. What should we expect from farmer field schools? A Sri Lanka case study. *World Development* 33(10): 1705–1720.

Umali, D. L., and L. Schwartz. 1994. *Public and private agricultural extension: Beyond traditional frontiers.* Washington, D.C.: The World Bank.

Van den Berg, H., and J. Jiggins. 2007. The impacts of farmer field schools in relation to integrated pest management. *World Development* 35(4): 663–686.

messages" and duplication, extension services have sometimes practised information "gatekeeping" and "thinking for others". At other times, extension services seem to have concentrated on the more visible and affluent farmers and farmers' groups. "Forestry projects have paid insufficient attention to the social barriers impeding the flow of information. The frequent omission of concerted efforts to contact women directly is ... glaring". This has amounted to the quasi-censorship of information and the preselection of options.

More information, not less, is required. Extension services must maximise the options and information available, providing a range of technological and institutional options on topics of interest to local users. Rural people themselves must be permitted and even encouraged to sort out and judge potential conflicts or duplications.

The ways in which extension services can disseminate information vary. Certainly, innovative and more extensive use can be made of the mass media and, in some cases, electronic networks such as the Internet. "If there is a common thread in the developments in the information industries over the past 15 years, it has been the erosion of monopolies and the enlargement of choice.... [Electronic networks] have multiplied the number of ways of receiving, exchanging and using information ... and will continue to do so". Extension services should also provide and promote decentralised fora for information sharing and dissemination. Of course, there is still no substitute for direct, personal contact.

Comparative Analysis and Evaluation

Surprisingly little research seems to be have been done or is under way on extension in general and forestry extension in particular. A search of FAO's database revealed no more than five activities even peripherally related to forestry extension. Progress in making extension more appropriate and efficient will be difficult without better recognition that extension is not only "researchable" but merits significant priority. More rigorous comparative studies, such as the one by Sen which statistically compares group, individual and whole community approaches to extension in Nepal, are desperately needed.

National extension services have a key role in promoting situation analysis, methodology development and comparative studies of extension approaches. But first, along with awareness of the needs and possibilities of research on extension, the extension process should be viewed and carried out as a potential learning experience, and therefore planned as such. In other words, systematic self-evaluation and self-criticism of the process of extension and its effectiveness should be integrated into the regular functioning of the extension services as well as into the daily lives of extension agents. In addition to being practical and concrete, extension research will need to be collaborative and cooperative, participatory and interactive. Those participating in it—foresters, farmers, extension agents, scientists, administrators—should accept not only the need for learning but, perhaps more important, the need for unlearning.

Involvement and Participation

There is a growing body of both theoretical and empirical evidence that the participation of local people and local institutions is necessary for effective extension and sustainable forest management and development: "Participatory extension methods are more likely not only to articulate farmers' knowledge and concerns, but also to facilitate the process of action research needed to identify and refine appropriate technologies and also help people develop their own problem-solving skills. To achieve this, simple methodologies are needed which enable extension staff to work with local people in helping them identify project design issues and indicators.

Consequently, there has been significant development of a "participatory toolbox"—tools, techniques and methods that aim to increase participation and involvement in the extension process, including diagnosis, testing and dissemination. Some questions about participatory methods remain, however. There appear to be cases where participatory tools have become ends in themselves and the overall objectives have been lost from sight. Extension materials have been developed, slide shows, drawings, audiocassettes, etc., but their fit into the local situation or the overall development goals and their impact have not been sufficiently considered. Good tools and techniques can make or break an extension system but, in themselves, they are not sufficient for success. Other questions about participatory approaches include the cost-effectiveness of certain techniques which seem "labour- and skill-intensive" and time consuming; the institutional rigour and accountability with which these approaches can be implemented; and the overall coherence of the approach and the reliance on the almost intangible elements of judgement and attitude.

There is also a growing consensus that extension has to be customised to the local situation; "a generalised extension approach conveying uniform recommendations for all may not suit any particular group of clients". Extension services, to a large extent, must be "demand driven" and responsive to the "clients'" concerns. There are a number of broad approaches to extension which relate to forestry. For instance, FAO outlines eight, each with its advantages and disadvantages, requiring adaptation to a well-analysed local situation. The involvement and participation of the clients in this process of "customisation" is essential.

Increased Professionalism

In many cases, forestry extension is a "new" field and has a different history from other sectors. The number of universities, for example, offering master's degrees in forestry extension is both limited and quite recent compared with agriculture. The number of forestry extension associations or professional groups is even more limited. Consequently, to respond to the changing world of forestry and to internal critiques, forestry extension will have to make major strides in improving the performance and the professionalism of extension staff. Three pertinent points are presented below.

Extension agents are often isolated from their colleagues in other parts of the country, let alone in other countries. A lack of contact with and exposure to others working with participatory extension methods means that an agent has little to measure himself or herself against and little feedback on performance. At the international, national and regional levels, the promotion of professional networks which sponsor exchanges of information, knowledge and experience would help create an esprit de corps and a professionalism among extensionists.

Another necessity is improved training and education. In describing the forestry legislation in Nepal, Pardo concludes: "... the forestry department must strive for new levels of professional performance. It must develop skills in working with local people as extension agents on the one hand, and manage forest ecosystems scientifically on the other." Extension training to develop extension attitudes, skills and judgement requires not only significant shifts in curricula in some places but also the use of adult training techniques, large amounts of practice and, in some cases, a type of apprenticeship where young and inexperienced extension agents work with experienced agents in real field conditions. Extension has to be integrated into the standard training of foresters while, for graduate foresters, systematic, ongoing, inservice training should be provided.

Finally, to encourage professionalism, career paths have to be clarified and systems of reward and accountability developed. As long as inappropriate work goes unscrutinised and good work goes unrewarded, there is little chance of insiders attempting to improve and outsiders giving much credence to the extension system. Systems of accountability must include collaborative evaluation of extension agents by their "clients".

EXTENSION WORKERS

The farmers had struggled for many years. Owing to increasing population pressure, land had become scarce. Consequently, they had to shorten fallow periods, as no suitable forest land was left for conversion to agricultural production. The farmers knew very well that rice yields would decline as a result—and yields had declined dramatically over the most recent years. Not knowing what to do, the farmers organised a village meeting and agreed to approach the agricultural extension officer to ask his advice. The extension officer who frequently paid a visit to the village listened to the farmers and promised to approach the research branch of the agricultural extension unit to explain the farmers' concerns. On contacting the researchers, he learned that the investigation of the problem of decreasing crop productivity had already been initiated. Researchers were in the process of editing the final research report and were confident that they had developed the appropriate technology for introducing sustainable farming practices. Before the start of the next growing season, the extension worker was able to return to the village with an innovative package, soil and water conservation technologies consisting of agroforestry components. The farmers were very grateful and immediately implemented the researchers' recommendations. They eagerly adopted the innovative technology offered. As a result of the timely research efforts and the efficient and effective communication, rice yields stabilised and even increased in only three years.

What used to be the farmers' biggest problem had been solved by research and extension. In fact the agroforestry technology developed killed two birds with one stone. Not only did it contribute to stabilising soil productivity, it also provided additional benefits such as fodder for livestock and fuelwood. Therefore, it was also appreciated by the Forest Department which had been concerned for many years about the high rate of deforestation and forest degradation through overgrasing and overcutting for fuelwood.

What may sound like a fairy tale is in fact a fairy tale. It is based on the paradigm of the conventional research process and our implicit thinking about the adoption of research results by clients or managers, in this case the small-scale farmer or shifting cultivator. According to conventional wisdom, clients have a problem which can only be solved by analytical researchers. It is the researchers' task to identify and analyse problems and to develop solutions which can be readily transferred from the experiment station to the field. Thus, researchers tell farmers what is best for the farming community and the environment. The link between the researcher and client is the extension worker whose role it is to assist clients in applying technology and to support the adoption process.

In recent years, soil conservation technologies for agricultural lands have received substantial attention as a way of increasing production and thus reducing the pressure to convert additional forest lands. Research and extension have offered farmers agricultural innovations that have been advocated as "sustainable" farming technologies with on-site and off-site benefits. However, adoption rates have been disappointing. The following discussion considers why extension has been less successful in resource conservation activities than one would hope.

The literature on land degradation and resource conservation in Thailand is filled with warnings about the consequences of deforestation; floods, droughts, loss of soil fertility, climatic changes, water quality reduction and sedimentation of reservoirs are referred to in practically every publication. To limit deforestation, research activities are directed at developing means that enable farmers to abandon "destructive" and "wasteful" shifting cultivation. On-site benefits and off-site benefits are expected to result from stabilised land-use systems. However, in spite of research, extension, incentives offered and some limited success, the sustainability of projects initiated in northern Thailand is being questioned. The soil and water conservation activities of the Thai-German Highland Development Programme (TG-HDP) may serve as an example.

Within the framework of the TG-HDP, soil and water conservation practices were developed and extended to farmers starting in 1987 with the objective of reducing erosion to acceptable levels, thus enabling more permanent crop cultivation. The practices consisted of contour buffers, alternating grass (*Brachiaria ruziensis*) and later perennial strips (hedgerows of leucaena and pigeon pea) with crops planted parallel to contour lines. During the first four years of project implementation (during which incentives were provided for farmers and extension workers) the number of farmers adopting the promoted technologies increased at a steady rate, but when the general incentive scheme terminated in 1991, there was an immediate and sustained drop in interest in soil and water conservation technologies. Many of the farmers interviewed by one of the authors in 1990/91 and again in 1994/95 expressed disillusionment with the recommended soil and water conservation practices.

When asked why they discontinued using the technologies they had initially adopted, farmers reported that the grass they had been encouraged to plant as an erosion control spread quickly into their fields and required increased labour inputs during the time of the year when labour was already in short supply. Weeding has long been recognised as an important determinant of crop yields in the tropics. The soil and water conservation technologies did not address this issue at all; in fact they exacerbated the weed problem. Therefore, it is not surprising that some villagers were angry about or afraid of the grass. Perhaps the most pertinent summary of the agricultural situation was made by one farmer:

> "Now there are too many people and the government does not allow us to cut any more trees to make new fields. Life was much easier for the older generation. But grass is certainly not the solution to the problems we are facing."

Population growth, opium suppression, the restrictions on shifting cultivation and the opening of the highlands to the lowland economy have produced a new set of needs, particularly the need for cash. Villagers were therefore more interested in alternative income-generating activities rather than attempts to improve subsistence production.

The authors argue that the main reasons for the disappointing results are to be found in the fact that the research did not address the right questions, i.e. those of the farmers.

Moreover, we would argue that the extension process aggravated the situation. A number of observations can be made to support this position:

- Information on environmental impacts is often confused, and presumed cause-effect relationships are uncritically accepted even in the absence of scientific data. For example, it is common to hear affirmations such as "the diminishing fallow period, declining crop yields, increasing rice deficiency, poverty, soil erosion and deforestation will accelerate and lead to ecological and social disaster in the north, unless changes are made". Such descriptions, although limited in scientific value, cry out for intervention; they are appealing to researchers, donor agencies and the international community, but perhaps not to the beneficiaries who do not perceive themselves to be sitting on top of a virtual time bomb.
- The perceived beneficiaries of research are often seen as part of the problem. They are viewed as conservative, reluctant to change and ignorant of the massive degradation that surrounds them. Viewed as helpless agents, their input in problem identification, analysis and solution is not sought. This is particularly the case where rigorous quantitative research is proposed. As a result, research is supply- instead of demand-driven.
- Researchers are frequently unaware of the local conditions, diverse circumstances, opportunities and constraints to which farmers respond.
- The researchers' overriding concern is to control runoff and soil erosion and it is assumed that a reduction in soil erosion will automatically result in yield increases. With regard to grain yields, different treatments frequently show—at least for researchers—unexpected results.
- Control plots labelled as "farmers' practices" or "traditional practices" often assume incorrectly that local resource users have fixed, inflexible practices, an assumption that is even more erroneous in the face of productivity declines.

The reader may have observed that the previous discussion hardly referred to extension. This is not an accidental omission because the only role extension workers played in the case described is that of uni-directional messengers carrying messages of technologies from the researchers to the target villages. Thus, messages, which from the perspective of many farmers had a questionable technical value, flowed one way only. In terms of results, fallow periods continue to be shortened and soil and water conservation as such remains a low priority for farmers.

From Messenger to Facilitor

The situation described in the Thai example is not unique. The same principles and dilemmas underlie the general approach to research and extension in many other countries. Many recent case-studies highlight the urgent need for a change in the role and attitudes of researchers and extensionists. In recent years, however, various alternative approaches have been initiated by governmental and non-governmental institutions. In Zimbabwe, for example, pilot activities with new approaches based on active farmer participation in research and extension have shown potential for lowering the barriers to the application of agricultural and forestry research results, increasing adoption rates and, thus, improving natural resource management and food security.

For example, the Conservation Tillage Project is being implemented by the Zimbabwe Department of Agricultural, Technical and Extension Service (AGRITEX), with support from the German Agency for Technical Cooperation (GTZ). The project started in 1988 with the aim of developing and testing conservation tillage systems. Over the course of the project, the elaboration of the research agenda was increasingly dominated by farmers and, as a result, the activities spread to other topics, including agronomic, biological and mechanical soil and water conservation techniques, agroforestry and other topics perceived as relevant by the farmers. The focus of the project turned strongly from research towards extension.

Other participatory approaches being developed in Zimbabwe are community-level planning and development, geared towards community-based resource management and development and implemented by an integrated rural development programme; and the participation extension approach of the ITDG Food Security project. All three projects are located in the semi-arid Masvingo Province. They work closely together and use the same philosophy and similar tools, thereby strengthening the impact and fomenting the institutionalisation of these approaches within the government extension service.

A participatory approach was developed and tested with individual farmers and communities. Incorporated in the approach are shifts in levels of decision-making and the recognition that the roles of the individual actors in conservation strategy development need to be redefined. Development is perceived more broadly and includes skill and socio-organisational development. In addition, incentive strategies are implemented only when needed and are not viewed as an automatic ingredient of every activity and project.

Participatory Innovation Development

The concept of participatory innovation development and extension is based on dialogue, farmer experimentation and strengthening of the organisational capacities of rural communities. Active farmer participation is the mainstay of the approach. This should not be mistaken for farmer participation in externally initiated activities (e.g. in supply-driven research projects); rather it means farmers identify problems themselves and subsequently initiate activities. They may call in support from other relevant actors, for example researchers, if necessary. A major and frequently underestimated result of the approach is an increase in farmer confidence. For the majority of the farmers, this is a precondition for becoming more innovative and organising themselves better for more effective natural resource conservation and development.

Stimulation of farmer experimentation proves to be a useful element in combining new techniques with appropriate traditional technologies. It increases the ability to assess options and to develop alternatives appropriate for specific ecological, economic and socio-cultural environments. Ultimately, the aim is to transform a technology-oriented research and extension approach into an output-oriented system leading to sustainable resource use.

Leadership training and facilitating dialogue and communication in village workshops are elements which have shown great potential for improving cooperation between all involved in natural resource management, the sharing of knowledge and the participation of all gender and age groups in extension and rural development. Not to be under-estimated is also the positive effect of strengthening local institutions and building up confidence in farmer-to-farmer extension.

The experiences and observations from Zimbabwe, particularly regarding leadership and cooperation, indicate that making headway in natural resource conservation requires a philosophical framework for the participatory development process that includes but goes beyond the adoption of participatory rapid appraisal (PRA) tools. This approach was introduced in the form of training for transformation (TFT), a training programme based on Freire's work and further developed by a local non-governmental organisation. TFT is based on raising awareness through participatory, dialogue-based education and it aims at empowering local people for self-reliant development. To implement its principles in meetings and workshops it offers a range of tools such as codes, role plays and poems. Elements of TFT are complemented by PRAs, diagnostic surveys and goal-oriented planning methods.

Role and Approach of Extension Workers

The traditional role of extension workers was clear-cut and straightforward. As messengers they provided the link between research and the client. Many perceived themselves to be in a teacher/student or, worse, in a teacher/child relationship. A participatory approach needs more than a messenger or a teacher. That the conventional one-way flow of information in many ways explains the poor performance of agricultural and forestry projects is an accepted fact. But is it sufficient to replace it with a two-way flow in which the teachers have an additional group of students, i.e. the researchers, whom they inform why the meticulously designed technologies are of no interest to the beneficiaries? We would argue that the key to finding real solutions and developing successful innovations is a participatory process that focuses on local institutional strengthening, the identification of the needs and prioritisation. Messengers or teachers, therefore, are not sufficient; initiating, supporting and maintaining a process of change requires facilitators.

The new extension workers' or facilitators' role is to provide farmers with background knowledge and technological options to stimulate discussions and encourage farmers to experiment with options and ideas. Extension must promote the sharing of experiences among farmers and also between farmers and other actors concerned with natural resource and rural development. In this way extension agents are not messengers themselves but they facilitate the flow of messages. They improve communication among the social actors of the development interface by strengthening local institutions. Over time the facilitators' role will be redefined and/or gradually taken over by community leaders.

TROPICAL FOREST MANAGEMENT OPTIONS

Options for forest resource development are different for distinct social groups. The provision and exchange of information among resource users regarding these options must therefore be tailored to reflect this diversity. This chapdescribes two recent experiences in forestry extension that have accommodated diverse local perceptions and interests in Paragominas, a municipality in Brazil's eastern Amazonia. The first example shows how participatory research helped to quantify the use and value of non-wood forest products (NWFPs) to the local communities so as to justify the sustainable management of forest resources for non-wood benefits, thereby mitigating the need for continued agricultural encroachment and destructive harvesting for timber. In the second experience, extension activities demonstrated to local loggers and policy-makers the technical and economic advantages of sustainable timber harvesting in the hope of reducing the rate of conversion of forest to pastures which, in turn, were quickly degraded. The emerging challenge is to identify ways that forestry extension approaches serve the needs of distinct client groups with apparently conflicting requirements. To the extent that extension reflects diverse options for human intervention in the same ecosystem, it can improve societal distribution of benefits from local resource use.

Diversity and Sustainability

Since the United Nations Conference on Environment and Development in Rio de Janeiro in June 1992, attempts at sustainable forest resource utilisation and management have increasingly taken into consideration the conservation of biological diversity. The term biological diversity generally refers to the conservation of animal and plant species and genetic variability within species as well as ecosystems and habitats. However, an operational definition of diversity should also encompass human cultural diversity. Different peoples express their resource needs and relationship to the forest resource in different ways. Over the years, local groups have often developed an extensive knowledge of biological diversity and its uses. Forestry and conservation efforts in natural forest areas are increasingly drawing on this local knowledge in a participatory process.

A key concept in biological diversity conservation is the increasingly interdependent nature of the world. While this interdependence can increase access by people to beneficial goods and services at lower costs, it can also result in added pressure on the fragile natural resource base. The linking of forested regions to domestic markets that are themselves increasingly affected by competition as well as by increasingly accessible international markets poses a challenge to those concerned with the sustainability of tropical forest resources.

This challenge is particularly intense along the frontier of agricultural and timber expansion in eastern Amazonia of Brazil. Here, highway construction and mining activities that began in the 1960s attracted settlers from elsewhere in the country with the promise of new lands, gold and employment. When they arrived, they were thrust into contact with isolated indigenous groups and riverine populations whose low population densities, subsistence lifestyles and knowledge of soils, vegetation and wildlife, etc. had allowed a harmonious coexistence with dense tropical forests that represent "islands" of biological diversity in the Amazon regions. Such islands retained species that had disappeared elsewhere in the region during the period of extended drought, which converted much of the Amazon region to savannahs. However, they have now come under threat from "development", in the form of expanding timber extraction, migratory subsistence farming, cattle ranching and other anthropogenic forces.

Typically, the perception was that rural development required the wholesale replacement of forest land by other land-use options—basically crops or pasture. In the process, timber was extracted as a by-product of agropastoral expansion. Despite the presence of literally hundreds of hardwood timber species, loggers focused on harvesting a limited number of commercially desirable species such as mahogany and cedro. After harvest, ranchers planted the aggressive *colonião* grass, and burnt their pastures frequently to retard forest regeneration. Farmers grew exclusively white crops—rice and cassava—in slash-and-burn cultivation for subsistence and the market. Even forest product gatherers and harvesters gathered NWFPs from a limited number of species such as babassu palm as they moved to a cash economy. Moreover, indigenous groups were encouraged to convert the forests and savannahs they had managed for centuries into simplified cropland and pasture.

At the same time, in its effort to forestall deforestation, the environmental movement launched campaigns to "protect" the natural forest from use of any kind—an often extreme and untenable measure.

These patterns have begun to change, however, partly as the result of the international outcry over deforestation in the region and partly as a result of the search by local communities and commercial enterprises for alternatives that are more economically viable in the long term. Timber operators, criticised for forest degradation and wasteful logging practices, have responded by dramatically increasing the number of species they use and by testing innovative management techniques. Ranchers are trying to recuperate degraded pastures to avoid encroaching on remaining forests, and new settlers in the area now hope to find ways to overcome the short-term limitations that had led to continued deforestation and the associated cycle of poverty and recurring migration. They have organised grassroots cooperatives and established agroforestry systems to produce timber and fruit from native trees. Research institutions and extension agents are playing an increasingly supportive role in this transition.

The recent tendency to promote the planting or management of a number of perennial species and production systems that can be better integrated with the local ecology, rather than to focus on either restrictive ecosystem protection or conversion of natural forest to monoculture plantations or to non-forest land uses, reflects a significant change in the approach of agriculture and forest development policy-makers and technicians and environmentalists alike. Development agents of all kinds are seeking a broad continuum of environmentally sustainable use options. The challenge, however, is to find options that are mutually compatible among the actors with competing interests in the same forest lands.

This challenge became particularly acute in eastern Amazonia, where non-governmental organisations have begun pilot activities in forest land-use research and community intervention in the same municipality, Paragominas, in the southern area of Pará State. Representatives of both small farm settlements and large-scale timber extraction firms have tried to define alternative land-use models that are more protective of the forest resource base. These experiences suggest ways to help reconcile competing resource demands in the tropics.

Capturing Local Knowledge for Extension

Local knowledge of the environment and indigenous systems of resource management, previously ignored or thought archaic, are now recognised as being essential to an ecologically sound strategy for forest utilisation. Formerly, agronomic and forestry extension had relied on the introduction of extensive production systems and models. It has now become imperative to begin with a clear understanding of the local rationale for current land uses. This knowledge can only be obtained by working directly with local people over the long term, documenting existing practices and constraints as well as the reasons behind them, and promoting alternatives only when it becomes clear that they would be appropriate within these systems. Such initiatives must be equally sensitive to both local ecosystem resilience and culturally ascribed values.

A good example of this emerging approach is an effort under way on the Capim River, a remote tributary of the Amazon in Paragominas. In the past, families in these communities relied on a combination of subsistence agriculture, the extraction of NWFPs and hunting and fishing for their livelihood, and had only very limited contact with regional markets, located many hours down-stream. Recently, farmers have been induced to sell timber on the municipality's burgeoning wood market as a source of ready cash. Harvesting was done inefficiently and degraded areas were converted to pasture or simply left to regenerate as brushland. As a result, plant and animal NWFPs became increasingly scarce. An international team of researchers associated with the government agroforestry centre for the humid tropics (EMBRAPA-CPATU) was asked by the rural labour unions to assist them in finding ways to keep their forest resources from destruction.

The research team began by making a participatory inventory of fruit, game, fish and medicinal plants consumed by 45 households over the course of an entire year. In addition, the population ecology of four NWFP species selected by the community was exhaustively examined and tree distribution, regeneration and productivity were documented. In both cases, it was the community members rather than the scientists who helped identify study parameters and carried out the resource inventory and utilisation studies.

Walks along a 40 km forest trail network with community members led to the identification of the most widely utilised products. After training, community members mapped the location of each of 200 trees, and monitored fruiting phenology over a three-year period. Satellite images were employed to pinpoint individual farmers' fields and distinguish adjacent forests that had been exploited and degraded as well as remaining areas of primary forest.

The participating household members maintained daily registers of products they consumed (by weight for game, fish and fibre products, and by unit for medicinal plants and fruits). Rather than assess the value of products solely in monetary terms, their equivalence in a locally essential commodity, cassava meal (*farinha*), was used to evaluate the importance of these products to household income.

This information, such as the farinha equivalent of NWFP consumption, was integrated into extension programmes. Activities included small group discussions, workshops for the whole community, exchanges between villages, theatrical presentations and the production of illustrated booklets. The objective in all of these activities was to ensure an understanding of and control over data that had been generated locally for use by the community in its struggle to protect and manage its remaining forests and to demonstrate and quantify the great value of the forest for the survival of local communities.

The Capim River experience exemplifies efforts by forest-dwelling communities to define effective management systems that can provide increased benefits while ensuring the conservation of the resource base. The virtue of the extension methods employed was to stimulate community members to give free reign to their knowledge and imagination and begin more long-term planning regarding how best to use the resources they control. Sensitivity to local knowledge and spatial awareness is critical in the definition of tree resources and products that should become the focus of management and marketing trials. Of course, quantitative analyses of the potential value, both marketed and non-marketed, of the full range of forest products will be essential if the communities are to be able to make rational, long-term decisions about the options for forest resource use. And unless the value of managing, harvesting and marketing NWFPs can be increased beyond that of timber, it is unlikely that the communities will choose them as the primary focus of forest utilisation. In the interim, pressure continues to mount in parts of Paragominas that remain forested, as timber harvesters expand their activities westwards. Efforts to reduce the destructive effects of logging practices are the subject of the second case-study.

Sustainable Timber Harvesting

Over the past decade, the degradation and destruction of forest resources in other parts of Brazil has resulted in greater emphasis being put on the Amazon basin as a source of timber. Paragominas is now the centre of the most intensely logged area of Amazonia; more than 200 sawmills operate around the clock during the seven-month dry season. The loggers are working their way westwards, leaving a trail of logging roads, new pastures, denuded banks along the Capim River and degraded local watersheds.

Ranchers and loggers of Paragominas come either from southeastern Brazil, where the transformation from dense coastal forests to fields for cattle and cropland occurred decades or even centuries ago, or from the central savannahs where forests are non-existent. Hence, they often perceive the tropical rain forest not as a resource in itself that is worthy of conservation and management but as an obstacle, and they hasten to fell it and convert the land to other uses.

Moreover, they are often inefficient harvesters. Until recently, most loggers gave little thought to the possibility of returning to the same land one day to harvest again. Their aim was to extract the most valuable logs before the land was converted to pasture by burning. They made no effort to promote regeneration or to avoid degradation. The returns to loggers and to the landowner, especially when supported by government fiscal incentives, were enough to support the costs of pasture expansion.

Lately, however, global concern over tropical deforestation has begun to be reflected in natural resource policy in Brazil—and incentives that favoured the conversion of forest to pasture have been progressively removed. Since the late 1980s, researchers associated with the Institute for Man and the Environment of Amazonia (IMAZON) have worked with loggers in the search for sustainable techniques for tropical hardwood harvesting and utilisation. Foresters involved in IMAZON's Wood Project (Projeto Madeira) conducted a series of studies in Paragominas that led to a better understanding of the rationale behind the loggers' and ranchers' current behaviour as a basis to identify land-use alternatives. Instead of vilifying their current practices, the research team found them rational, given the mix of market incentives, technical information and equipment available in the region. As it was unable to address the wider issues of market incentives and land tenure, IMAZON embarked on a programme to improve the technical basis of forest management in the area.

By agreement with a landowner who was in the process of selling timber rights to loggers, IMAZON established a 50 ha research and demonstration plot for sustainable management techniques. Measures included modification of the silvicultural system, including a reduction of the rotation period from 90 to 30 years, the initial marking of trees to be cut, the removal of vines that would otherwise have pulled down or broken neighbouring trees, the use of wedges to control the direction of felling, and site planning to minimise road extension and damage. Compact skidders were employed in preference to heavy caterpillar tractors to avoid understorey damage and soil compaction. The additional effort involved in undertaking such management methods was carefully monitored.

The initial costs in the demonstration plot were slightly higher than those incurred on a neighbouring plot which simultaneously underwent traditional logging, and the harvested wood volume taken out of the forest was less than that removed under the traditional practices. However, under the traditional practices, with the higher volume of harvesting and significantly more damage to the remaining trees, a 90-year rotation cycle was necessary to allow the forest to restore itself. Under the improved practices, which had a 30-year rotation, the present value of site output increased significantly.

Furthermore, were current logging practices to continue, forests in Paragominas would be exhausted by the end of the next decade, resulting in a declining tax base and collapse in the local economy unless other non-forest sustainable systems could be developed. Under sustainable management, the forest could continue to provide raw materials indefinitely, thus providing continued employment, income and public revenues, as opposed to the current "boom and bust" scenario.

Once these studies had been completed, IMAZON began a series of extension activities designed to communicate the results to loggers and other groups. During the felling season, an on-site demonstration of managed extraction techniques was held with participation by managers of many local logging firms. With support from the World Wide Fund for Nature (WWF-Brazil), an extension video was produced and widely distributed for environmental education and training purposes. In the video, loggers express their enthusiasm to explore less aggressive, more sustainable management techniques, and opt for a greater utilisation of timber that was formerly considered waste for furniture manufacturing so as to add value and increase local employment.

An extension manual describing appropriate practices is being prepared by IMAZON for wide distribution and use by state and federal agencies. The manual contains a financial analysis suggesting that those who adopt the sustainable system will have higher initial costs but will be able to reduce the forest land area they must manage in the long term for the same value of timber production, thus reducing land and transport costs.

FORESTRY EXTENSION WITH RURAL WOMEN

This section discusses the role of women in forestry extension in Pakistan. For instance, in Pakistan, purdah, the social segregation of men and women, is the cultural norm. The government forest service in general and the extension service in particular are exclusively staffed by men. Moreover, development efforts including social or community forestry projects have not yet developed approaches that are culturally compatible with the participation of women. This has resulted in the virtual exclusion of women from forestry and watershed management development efforts, despite the fact that women are responsible for many activities that have a direct bearing on natural resource conservation and use.

A Brahui woman works an average of 17 hours each day during the productive season. Gender analysis of women's routines reveals that from April to October, most women in Noza spend over 60 percent of their time in productive work and the rest of their time is divided between domestic chores and social obligations.

According to participants from Balochistan in the First National Conference of Peasant Women, held in Lahore in 1991, "our work includes domestic work like cooking, washing clothes, fetching water and fuelwood and making dung cakes. Our work also includes looking after farm animals, cutting grass and cleaning up the place of the animals, giving them fodder and water and finally milking them. In the fields we tend the wheat crop—pulling out onions and putting them into sacks—weed the fields, pick and cut vegetables and then take meals to the men working in the fields. Not only do we do this, some of our women also do sewing and wage labour."

Although Brahui women and children use natural resources and significantly affect their sustainability, they have little or no say in the decision-making process. For example, wells and land are owned by men. In some tribes, women are not allowed to plant trees, especially fruit-trees, because it is believed that if a woman plants a tree, the tree will wither and die or it will be barren and not produce fruit.

Since custom dictates that women must be separated from men, it is necessary to train and prepare two teams of extension agents to do field work. Although these teams of men and women cannot work directly together, they are jointly responsible for integrating gender issues in overall village planning and development.

In Balochistan, the Forestry and Wildlife Department has not as a rule employed professional women or women forestry extension agents. Women in Balochistan who pursue professional careers tend to go into medicine or education, professions that are traditionally more acceptable for women. There are almost no role models of women working in forestry, agriculture or with livestock and the problem is compounded by the few positions reserved for women in the provincial government overall. Moreover, for the few professional women who are in government service, it is much more acceptable and desirable to work in an office than to go out into villages. Families will only allow women members to travel to rural areas in controlled conditions and then only in project or department vehicles.

To reach women, the Pakistan pilot project has had to hire women "group promoters" from outside the Forestry and Wildlife Department. Instead of advertising in the province for candidates with advanced degrees, who would come from urban centres such as the provincial capital, Quetta, local Brahui-speaking women without advanced degrees were hired as group promoters. Project management upgraded the skills of these women group promoters to ensure professional equity between men and women working under the project.

For women group promoters, learning to use participatory approaches requires putting aside considerations of social status and conventional attitudes while learning to listen to and cultivate an "openness" to village women. Changing attitudes is a difficult and complicated process, and the project drew on experiences by trainers in Shirkat Gah, a non-governmental organisation (NGO) working for the empowerment of women in Sindh Province. While training women group promoters to work in urban low-income areas in the city of Karachi, these trainers encountered resistance from trainees that seems to be common in Pakistan. Although the women being trained were "themselves middle-class, they appeared to resist identifying with low-income commu-nities. Symbols of an 'educated middle-class' status were evident in their cloth-ing, and the carrying of large bags and files on field visits emphasized differ-ences and perhaps reflected a fear of poverty and rigid class consciousness". The project had to emphasize the need not to make these class distinctions.

Critical to reaching rural women is the ability to organise meetings. However, the target group, the village women in Noza, had no experience of holding meetings or doing exercises as a group. Most of them had never been visited by female group promoters or government extension workers before the advent of the project. "First meetings are always noisy and chaotic, with everyone speaking at once, babies crying and children crowding into the room to see what's happening". In one village, young boys on nearby rooftops started pelting stones at women meeting with project staff because the women tried to keep them out of the meetings. The next visit, group promoters brought footballs to the village, keeping the boys busy and allowing the women to meet in peace. The footballs were an effective way to buy time until the women's meetings were no longer perceived to be an unusual occurrence.

The first step in the process of involving women in natural resource management was to conduct a participatory rural appraisal (PRA) with the village women. This was conceptually difficult, as the traditional hierarchy and culture provides individuals and groups with a stylised forum for communicating needs to traditional leaders and to outsiders such as project staff. PRA, on the other hand, depends on people speaking for themselves. A good deal of time and effort was devoted to developing interactive tools for a PRA that could be utilised by the women, nearly all of whom were illiterate.

One productive PRA exercise was the preparation of a map of the village. It is noteworthy that this activity drew a good response at least in part owing to the "privilege" of using pens and paper, usually restricted to the world of the "literate"—in this case the exclusive domain of a few village men. A particularly interesting result of the mapping exercise was that the women, in contrast to village men who participated in similar exercises, consistently include natural resources in the village map. They consistently draw the wells, the fields and the uplands, reflecting the importance of these areas. This confirms that "women's perspective about their environment is often more multidimensional and multifunctional than that of other members of society", an element crucial to the sustainable management of the watershed.

Another PRA tool adapted for use by the village women was the "daily time profile". This tool was used to promote discussions of women's daily routines and seasonal activities. Given that the seclusion of women within the family compound is associated with a higher economic standing, women are often reluctant to speak about tasks they perform outside the home. Village women experienced difficulty in listing all the different activities they carry out, so the group promoters designed a series of colour pictures depicting what rural Brahui women might do during a day. In fact, the village women found it easier to talk about their daily routines when holding the pictures. If women identified additional activities that were not represented by the pictures, new images were drawn up by the group promoter.

After identifying the various activities they undertook, the next step was to calculate how much time was spent on each. At first, this was difficult for the women. They do not measure time in hours, but traditionally by counting the sections of the day. To simplify the exercise, group promoters photocopied the time profile pictures in three different sizes to get a better comparative approximation of time spent on various activities. Although they were still able to participate in the exercise and this interactive tool allowed women to create a visual documentation of their daily lives, the transformation of the images from colour to black and white in the photocopying process made it significantly harder for the village women to interpret and use the drawings.

The priority needs identified by women in Noza through the PRA focused on infrastructure development (clinics, schools, etc.) and not on conserving natural resources. Given their responsibility for a wide variety of labour-intensive household and village tasks, and given that they are excluded from the decision-making process concerning natural resources, this seemed logical. Like rural women in many parts of the world, they want access to piped water in the village, they want access to schools and health facilities and they want to increase their family income. However, a pilot project for upland conservation and development can seldom justify the establishment of health facilities or schools. On the other hand, income-generating activities help build the skills, organisations, etc. necessary for long-term sustainable management of natural resources. Therefore, project entry point activities with women have centred on income generation.

Promoting Strategies

The project strategy was to develop income-generating "packages" while, considering cultural norms, looking for ways of involving women in managing the environment. Project staff assisted women to create self-help organisations through which they could learn to hold meetings, make group decisions and manage credit. Women need to build up their self-confidence and ability to express themselves. They need to be able to discuss ideas and to have access to information. Organisations provide them with a forum for developing these skills and for receiving training in other skills. At first it was difficult for the women to decide what they wanted to do because they have had little exposure to most income-generating activities and they are not involved in marketing. After doing a simple feasibility study, women select income-generating activities such as raising chickens for egg production, lamb fattening or tailoring. A women's organisation was built around these activities. In each case, an activity is selected by the group and group decisions are recorded by group promoters until a local community member who can keep written documentation for the organisation is identified. If the village is large, two or three women's organisations are created. Women are also asked to agree to start a savings programme where each member contributes to a cash box on a monthly basis until a minimum amount is reached and a bank account can be opened.

Office bearers such as the chairperson, deputy chairperson and treasurer are elected by the group for a one-year term. The group also gives the organisation a name. A spontaneous competition has emerged in Noza to find the most original name of a plant or wild flower growing in the watershed, so organisations have names such as Wild Tulip, Wild Iris or Wild Almond Blossom.

Initially, group promoters try to hold two well-planned and interesting meetings a month with each organisation. Village women often dress up on meeting days, which is a sign of the importance accorded to them. Attendance is checked to stress that regular participation is important and that decisions cannot be made unless all members are present and a consensus is reached. Group office bearers also learn to record monthly savings in members' passbooks. Although meetings are geared around income-generating activities, the purpose is to train women in leadership and managerial skills so that they can eventually manage their organisations independently and play a bigger role in natural resource management.

Training Courses

All project activities are supported by short training courses. All participants are invited to attend training and, after an initial hesitation, women seem particularly to enjoy and benefit from the courses. The courses are designed to maximise participation and often involve the use of picture cards, group discussions and practical demonstrations. Children and young girls are encouraged to take part in training with their mothers.

The project's policy has been to contract women extension agents and trainers from Mastung District when possible rather than to rely on urban areas. It is easier to find women extension health workers in Mastung District. For example, village women wanted to construct pit latrines so the project contracted a Lady Health Visitor to give basic hygiene training. She used existing training materials produced by UNICEF and adapted to the project area. However, all extension agents required a financial incentive from the project because working under a project in rural areas was considered to be outside their normal departmental duties.

Another problem related to the organisation of training courses was that cultural norms dictate that women must not travel without another woman. Therefore, since a group promoter always has to accompany a trainer, the project had to hire an additional vehicle for transport. Moreover, project staff have had to prepare the training courses for chicken raising, egg marketing and lamb fattening because there were no women extension agents with this experience. The creation of an organisation is an advantage because it is easier to introduce staff from government agencies or NGOs to groups of women willing and ready to learn. For example, women from an agricultural office in Quetta, experts in "jam making" who had never been in villages before, could come to organisations in several villages where the women wanted to learn how to use excess fruit from the orchards. As women's self-confidence grew, they started taking the initiative and solving their own problems. The first women who took part in chicken raising for egg production from the Jasmine organisation were keen to manage their own chickens and not to get their chickens mixed up with others in their compounds. Their solution: they took the initiative to dye the chickens vivid colours like cobalt blue, chartreuse and magenta.

Exposure Trips

Early in the project, the practice of taking women on exposure trips and exchanges was initiated. Rural women are often isolated and cut off from information. Through exposure trips they have an opportunity to see and learn and exchange ideas. For this purpose, the project hires a van or a bus for women only and group promoters accompany them on short day trips. Women must have permission from their husbands or families to leave the village and, at first, only the old women were allowed out with the project, but gradually more and more women of different ages were granted permission to travel.

The first exposure trip brought 14 women from three villages to visit demonstration plots of "salt bush" (*Atriplex* sp.) at the Arid Zone Research Institute (AZRI) in Quetta. As a result, one woman convinced men in her village to give a plot of land to be planted with salt bush in cooperation with the Forestry Department. Other trips have been made to the well-known annual Sibi livestock fair. On one occasion, the hired bus broke down in the Bolan pass and some of the women did not reach home until after midnight. That women were allowed to participate on subsequent trips was an indication of the level of trust that had been established with project staff.

Exposure trips have also been organised to bring together the leaders of the newly created local women's organisations. During the visits, the women find themselves in the position of having to describe their organisations and activities to other persons who, although women, are strangers. This strengthens their ability to make concise, sequential presentations, think logically and comprehensively about their activities and develop a professional capacity for leadership. For example, Zer Bibi, who is deputy chairperson of the "Almond Blossom" women's organisation in Raza Mohammad village, recently visited the chairperson of a Balochistan Rural Support Programme-assisted organisation in Dasht. Zer Bibi had to explain how Almond Blossom organisation worked and what major activities had been carried out in the village through the organisation.

In turn, with this experience, Zer Bibi was able to provide a model for other village officers to follow at a Village Area Convention for farmers and their wives, organised by the project in 1995. Nineteen village women from six villages attended this convention. There were only five women from a single village who attended the first project convention in 1993. Zer Bibi was asked to present the Almond Blossom organisation first during a general presentation by officers of organisations, and she therefore provided a model for other women to follow when making their presentations.

The women and the project are actively seeking ways to address the critical fuelwood situation. In one village in the project area, the mullah's wife has a solar cooking stove. She knows exactly how long it takes to boil water, cook meat or rice or lentils and even bread. Her husband will not give her permission to travel outside her village but he has allowed project staff to organise an exchange visit to her house for other village women who are interested in learning the advantages of a solar stove and how to use it. The fact that project staff can now organise a demonstration by a village woman of a solar cooking stove, which will highlight fuel conservation in Noza, to a group of women from the area illustrates the progress that has been made in facilitating the participation of women in their local development and in natural resource conservation.

The project has begun to involve village women in the planning and implementation of tree planting in compounds—although the trees are actually planted by children to avoid conflicting with the social taboos. This activity has been carried out at the same time that village men are planting trees along water channels and in some uplands. In three villages, the project has persuaded a husband to permit his wife to take care of demonstrations of dwarf apple trees on their household land—a significant accomplishment. One of these women has planted vegetables between the rows of trees and now holds a multipurpose demonstration.

In the most recent phase of the project, strategies have been adjusted. As new villages are involved, one of the first activities is to draw a map of how villagers would like to regenerate common lands or uplands with assistance from the Forestry Department. Village women have been asked to decide which shrub species they would like for fuel and which medicinal plants they would grow on the uplands. Conversations have been tape-recorded so that local knowledge about medicinal plants can be passed on to all the villages, although some women refuse to allow their voices to be taped. The Forestry Department has agreed to assist in developing a seed collection activity. Women will collect the seeds of preferred medicinal plants and sell them to the Forestry Department for planting on the uplands. Implementing this programme calls for closer cooperation between men's and women's field teams and between village men and women. In order to strengthen this cooperation, project staff are planning to do village-level "gender sensitisation" training later in the year with assistance from the Aurat Foundation.

The Balochistan Forestry Department appears satisfied with women's participation in the Upland Conservation and Development project but, in order to sustain this sort of action, it will be necessary for the department to create permanent posts for strengthening women's involvement in social forestry and management of the environment. It is not easy to convince policy-makers and provincial financial departments to create new posts for women in extension. Obviously the experience of a single project cannot provide the drawstrings that totally lift the curtain on ideal participation and integration of women in forest and watershed management. However, the Balochistan pilot project has demonstrated that progress, although fragile, can be made even under what might appear to be particularly difficult conditions. It may be useful to attempt to single out several lessons learned from these recent efforts within the framework of better integration of women in natural resource management:

- In situations where women are isolated and unable to work with men from outside their families, women must be hired and trained as extensionagents and group promoters if women resource users are to be reached;
- simple techniques and opportunities can be found to overcome the small hurdles and social taboos hindering the better organisation of women, allowing what was once seen as exceptional to become ordinary;
- keys to sustainable natural resource management by women include capacity and confidence building and the development of organisations and organisational skills.

In summary, the approach of extension for women in this situation differs from that for men. It appears that men's characteristics, including mobility, control over the means of production, education and general level of confidence, allow for an extension approach that is closer to a transfer of technology. At least until women gain many of these attributes, partially through the efforts of a group promotion extension approach, the transfer-of-technology approach will lack sustainability. It is necessary to consolidate the small advances so that what was unthinkable a few years ago becomes not only thinkable and possible but ordinary. This requires an approach that is respectful of cultural norms and of all members of society.

References

Chambers, R. 1983. *Rural development: putting the last first.* London, Longman.

Hoskins, M.W. 1987. Agroforestry and the social milieu. *In* H.A. Steppler & P.K.R. Nair, eds. *Agroforestry: a decade of development.* Nairobi, ICRAF.

Hunt, J. 1993. Extension. *In* L. Pancel, ed. *Tropical forestry handbook,* Vol. 2. Berlin, Springer Verlag.

Unasylva. 1984. "Forestry extension: making it work". *Unasylva,* (36)143.

6

Aquaculture Extension Systems

Agricultural extension is now recognised as a potent and critical force in the development process. It has been defined as "a service or system which assists farm people through educational procedures in improving farming methods and techniques, increasing production efficiency and income, bettering their levels of living, and lifting the social and educational standards of rural life."

Functions generally attributed to agricultural extension include "to communicate agricultural research findings and recommendations to farm people, and to bring farm people into contact with sources of practical and useful information through organised group action". Activities of extension involve "agents working with farmers to teach improved farming practices, new techniques, and more productive or more efficient technologies ... organised within a system with administrative direction and support."

The structure, goals, strategies, and organisation of any extension system vary with the multitude of sectoral goals and objectives of national governments. Therefore, any particular extension system reflects the underlying culture, philosophies, and priorities of the political, economic, and social environment in which it evolved.

PURPOSE OF EXTENSION IN AQUACULTURE

Aquaculture is a new technology in many parts of the world, while in others it is a well developed tradition. Such differences affect the scope and nature of extension efforts among different nations. The primary purpose of aquaculture extension is to improve production of aquatic organisms for food and wealth for farmers, as well as for the nation as a whole. Extension programmes are justified by the fact that economic development resulting from extension activities should more than compensate for the national investment required to provide these services.

Extension activities can be divided into three primary functions:

i) education,

ii) services, and

iii) liaison between government agencies, universities, and producers.

In some cases, aquaculture extension programmes include additional duties. Education is often thought of as the primary function of extension activities. Extension programmes are designed to teach producers new technologies developed through research, and can take many forms. For example, extension agents may hold formal workshops, conferences, or producers' meetings to present new information. Agents also work one-on-one with individual producers, groups of producers, or cooperatives.

Demonstration ponds may be developed on a particular farm to show to other farmers in the area the viability of a particular new aquaculture technology. Many aquaculture extension programmes provide services to farmers. Some provide production inputs, such as fingerlings or transportation for feed; others build ponds for farmers, and those which target subsistence or semi-subsistence farmers may provide credit. In some cases, these services may be quite technical and consist of disease and water quality diagnoses,

The third primary function of extension programmes is a role of liaison with other government and private institutions. Extension agents provide an important link between researchers and farmers. In most cases researchers are not adequately skilled in communicating with farmers and frequently do not have the regular contact with farmers needed to transfer technologies effectively. Extension agents provide this important link. Finally, agents often coordinate contacts with other government organisations which provide health, housing, credit, or other services to the farmers.

ROLE OF GOVERNMENT IN EXTENSION METHODOLOGIES

Government policies and philosophies determine the structure of extension activities, the type of activities undertaken, and identify the target group. For example, some countries place great emphasis on rural development and programmes designed to improve the standard of living of rural poor; others place primary emphasis on industrial development or large-scale agriculture to stimulate economic development which may provide significant benefits to society as a whole. Extension is therefore matched accordingly. At times this is not always easy.

One important source of funding for extension services in developing countries is international technical assistance from multilateral and bilateral donors. However, the criteria set by the philosophies and goals of the host country and donor agency are often in conflict. A nation which places emphasis on food security may put priority on projects designed to improve the quality of produce consumed by the farm family.

The value of products consumed in the home alone, however, is generally insufficient to satisfy donor requirements of acceptable financial rates of return. These varying philosophies of government on the process of development dictate which groups of people are targeted or affected by extension efforts. The manner in which extension is subsequently carried out depends consequently upon the characteristics of the target groups.

TARGET GROUPS

The selection of target groups for extension programmes often results in a highly emotional debate. Selection is generally determined by government policy, although government policy does not always represent the attitudes and philosophies of all individuals in society. Selection of target groups may precipitate debates which range from ideological/philosophical posturings to pragmatic assessments of a nation's capability of providing extension services dictated by a particular philosophy. Extension services represent a large investment with returns occurring over the long-term. Rarely are returns immediate in nature. Budgetary resources may not permit the level of investment which a sense of moral obligation to strive for social justice may require.

Who will be the beneficiaries of the aquacultural extension system? Target groups can be identified and categorised in several ways. In many countries policy debates are common on whether to carry out extension with the purpose of maintaining low consumer prices, or helping rural producers to achieve higher standards of living through higher prices for their commodities.

Aquaculture projects often specify goals of enhancing rural incomes, nutrition, and overall food security for smallholders. At the other end of the spectrum the need for foreign exchange has resulted in capital-intensive aquaculture systems for large-scale commercial producers.

A second policy issue revolves around emphasis on private producers or cooperatives. Often the argument is phrased as a decision to work primarily with large- or small-scale farmers. This decision is closely related to national philosophies on economic development strategies.

Private Producers

Some countries identify private producers as the targets for extension activities. Private producers can defined be defined as farmers who work for themselves, receive benefits from their management and control of the aquaculture enterprise, and assume the consequences of any wrong decision. Private producers can be small-scale farmers working individually, or large-scale commercialised farmers. The common denominators are the private ownership and management control of the farm pond(s). Private producers have strong incentives to adopt those technologies which will benefit them the most, and from which a tangible reward can be obtained for their efforts.

The challenge to extension, in general, is to demonstrate effectively to farmers the benefits of any proposed technology, and to ensure adequate feedback from farmers as to its on-farm feasibility. Private producers can be difficult to work with if farms are small, isolated, and difficult to reach. In these cases extension agents may invest enormous amounts of energy, fuel, and effort to visit remote areas with relatively little time to work directly with the farmer.

Ambitious programmes, which call for working and attending to large numbers of such farmers, may place unrealistic demands on both the energy of extension agents and transportation resources of many countries. Some countries have developed ingenious methods of handling these types of situations.

On the other hand many large producers provide a different set of problems for extension programmes. For example, large producers may be more educated than extension agents and consider them poorly trained, and some purposely try to avoid extension agents if they fear that the agent may communicate their production secrets to other farmers.

Despite these difficulties, the trend in aquaculture extension has been to move away from emphasis on group and cooperative efforts to services for private producers. Efforts to encourage community and group aquaculture projects have generally met with little success.

Cooperatives

Many development efforts of international assistance agencies have been oriented towards cooperative groups. In this approach individuals are not favoured with special privileges, and extension resources are used efficiently by working with a group of people rather than one person at a time. The investment required in both capital and labour to construct fish ponds has been the basis for encouraging cooperative or group efforts in aquaculture.

Instead of 10 small ponds constructed at high cost per water surface area, one large pond could be constructed at a much lower cost and would supply the same number of people with fish. Moreover, one large pond requires only one good site, while 10 small ponds require 10 good sites. Individual subsistence farmers may have insufficent land to set aside part for a fish pond, but the government may be willing to provide land to a group of farmers for an aquaculture project. However, grouping farmers is not always easy.

Cooperative efforts in Panama appeared to be hindered by the organisational complexity related to socio-economic inequality, existing cattle ranching operations, and divisions between members and non-members. But younger, married, and relatively well-informed men with large households were especially attracted to group fish culture projects.

Socio-economic homogeneity, and socially esteemed and influential leaders (but not rich) are factors which contribute to the success of group projects. Nevertheless, many cooperative groups have not functioned well. In general cooperatives are short-lived, but there have been exceptions. The basic problem is to provide adequate incentives for farmers to share the workload. Inevitably some individuals work harder than others, and become disgruntled when those who do not work as hard receive the same benefits as they do.

In fish production it is especially difficult to determine whether an individual has contributed his or her share of work, or if he or she has harvested more than a fair share. In cases where cooperatives have been successful, there is generally a strong and highly-respected leader who has been able to maintain unity. If this person leaves, however, the cooperative usually fails.

Subsistence Producers

A common debate in development philosophy related to extension is whether to work directly with the poorest people, or to work with wealthier farmers and hope that the technology "trickles down" to lower economic strata. In societies which are highly stratified in economic and social terms, choosing to work only with the wealthier class prevents the technology from reaching the poorer strata. In practice the "trickle down" theory rarely works.

In 1979 the World Conference on Agrarian Reform and Rural Development declared that "... most development efforts have not yet succeeded in satisfying the aspirations of peoples and their basic requirements consistent with principles of human dignity and international social justice and solidarity, especially in the rural areas of developing countries. Rural development strategies can realise their full potential only through the motivations, active involvement, and organisation at the grass-roots level of rural people, with special emphasis on the least advantaged."

Efforts directed towards the "poorest of the poor" meet with enormous obstacles. Many of the world's farmers are landless or tenants. Poverty itself indicates a very low level of resources to work with to provide both food and income for the family. It is far easier to work with people who have a certain level of resources which at least permits them to provide some amount of surplus production to accumulate capital or other resources. However, if the participants in a project are all extremely wealthy the technology may be perceived as one which is only appropriate for the upper economic strata.

In some cases the expressed target group may be subsistence farmers, while in fact the majority of extension efforts are devoted to larger-scale producers. If donor agencies insist on funding only projects devoted to subsistence producers then there is a strong incentive to state them as the target group whether or not this is based on reality.

RESPONSIBILITIES OF EXTENSION PERSONNEL

Education

The primary responsibility of extension personnel world-wide is that of education. Technology transfer is an educational process orchestrated by extension personnel through both formal and informal means. Formal education by extension personnel takes the form of workshops designed to focus on specific topics, conferences, seminars, and producer meetings.

Extension personnel also engage in in-service training of extension agents who are responsible for day-to-day contact with producers. These formal education sessions may focus on a wide array of different topics, from specific management practices to more general technical theories of relationships between stocking and harvest size, and even oxygen fluctuations in ponds and disease diagnosis. Extension personnel also spend a great deal of time in informal education through farm visits.

Farm visits also provide farmers with the opportunity to ask questions on different topics, problems, and issues which are not covered in formal seminars. Farm-level demonstrations are techniques of informal education in which the extension agent demonstrates a particular technique to improve farm operations for an individual or group of farmers. Hands-on training and education, working side by side with farmers, is an effective albeit time-consuming means to transfer technology.

Educational activities also include dissemination of information to producers and consumers, through, for example, newsletters, newspaper articles, television and radio broadcasts, and utilisation of other media. Specific technologies or techniques may also be presented at fairs, markets, and other public gatherings.

Printed materials in the form of popular magazines or comics, manuals, bulletins, fact sheets, and books, all contribute to the educational process. Almost all (87%) of the countries responding to the survey utilised formal education methods. Only two countries indicated that they did not have organised extension services for aquaculture. All but one country utilised informal education methods. Uruguay is a country with little aquaculture extension.

Diagnostic Services

Extension services for the agriculture sector invariably establish and provide technical services; for example, seed and soil testing services. These services may be free, charged at cost, or operated for a profit. Most aquaculture extension services provide water quality diagnostic services. Extension agents are often provided with a water-test kit as part of their standard equipment. Basic parameters for water quality can be analysed to identify its suitability for raising fish, or other aquatic organisms, and to determine the need and quantity of lime and fertiliser. Extension agents with water test kits also have the capability to provide advice on water quality problems related to pond fertility, oxygen, and management.

Many extension services provide disease diagnostic services and recommend treatments to aquaculture producers. This service becomes important in areas where aquaculture has reached industrial levels of production. Most disease and water quality problems originate in extremely fertile ponds as a result of high rates of feeding. In many countries the lack of production inputs for fish farming, such as feed and inorganic fertilisers, results in low pond fertility, and hence production. For most extensive systems, with low levels of production, disease problems are rarely significant and do not justify investment in a laboratory and equipment for disease diagnosis.

In rural areas transport and communications are such problems that diagnostic services are not practical. Most disease or water quality problems require a quick response. By the time the agent arrives at the pond, the problem and symptoms may be past. In such situations there is virtually no need for disease or even extensive water quality diagnostic services.

Supply of Production Inputs

Some countries provide production inputs to new and aspiring aquaculture producers. The decision to supply or not to supply these inputs is most often dictated by overall prevailing political philosophies. For example, in some countries development strategies are based on the concept that the government is responsible for providing all essential inputs. This is often the case where individual producers have extremely limited resources.

Farmers participating in development projects may have difficulty arranging transportation for production inputs. Extension agents often are assigned transportation, and use it to transport materials to farmers on regular visits. For example, feed for livestock (pigs and poultry) which produce manure to fertilise fish ponds may be transported by extension agents. Only five of the 15 countries responding transported production inputs to fish farmers.

The five were Rwanda, Tanzania, and Zimbabwe (in Africa), Thailand (Asia), and Panama (Latin America). In many of these countries transportation itself is often difficult, and fuel shortages are commonplace, particularly for the government sector. These factors drastically restrict the functioning of government programmes. Six of the 15 countries responding (Cameroon, Rwanda, Zimbabwe, Thailand, Colombia, and Guatemala) indicated that, as a rule, production inputs were supplied to fish farmers. Three of these (Cameroon, Zimbabwe, and Guatemala) specified that the only input was fish seed.

Fingerling production requires a higher level of management than grow-out. New fish farmers lack the required level of management skills to produce their own fingerlings successfully and reliably. Fingerlings may also represent a substantial cost for subsistence or semi-subsistence farmers. Many nations have developed a policy of supplying fingerlings to producers participating in government projects.

The rationale involves a somewhat paternalistic attitude that poor farmers cannot afford to purchase their own seed, and that on-farm seed production is difficult to achieve. Seed distribution is also a quantifiable means of justifying a government station, personnel, and other facilities, and therefore the need for fish seed may be used as a tool by politicians to lobby for hatcheries in their home regions. The pressure to justify the existence of hatcheries may also result in production of "paper" or "book" fish, instead of accurate accounts of fingerlings harvested and distributed.

If government hatcheries consistently supply seed to producers at no charge, the producers may become dependent on the government. Free seed from the government may prevent development of a fingerling production industry and may hinder the growth of a viable aquaculture sector. Panama has an explicit policy of selling fingerlings at production cost, and encouraging farmers to produce their own seed on farm.

Aquaculture development experts believe that the first step towards a viable sector is to transfer seed production technology to farmers who possess adequate resources and conditions. In Liberia government hatcheries have been transferred to the private sector and are operated by private entrepreneurs. While government capital was used originally to construct the seed production infrastructure, government management preferred the greater efficiency afforded by private producers. The same policy has been practised by many developed countries.

Enforcing Regulations

In some countries, particularly those in which aquaculture functions are included in a ministry responsible for fisheries, natural resources, and wildlife, aquaculture extension responsibilities may be assigned to agents already charged with regulatory and other policing activities. Extension agents have responsibility for regulation and enforcement of laws governing fisheries and natural resources. In the three Latin American countries aquaculture extension is carried out by a fisheries ministry or institute, whereas Zimbabwe has both extension and natural resource regulatory functions in the natural resource ministry. There were, however, several of the other countries in which a fisheries or natural resource ministry was responsible for aquaculture extension, but the extension agents did not have regulatory responsibilities.

The rationale for including regulatory responsibilities with those of technology transfer stems from the limited resources available to many government agencies, and the need to economise. If any government agent is in the field, and particularly working with fish, it is more cost effective to have the agent patrol fisheries waters and look for violators of fishing regulations. Assigning regulatory responsibilities to an extension agent, however, automatically relegates the agent to a position of antagonist to the farmer. This is unfortunate, as an effective extension agent is one who gains the trust and confidence of producers. An effective extension agent is perceived by client producers as having their best interests at heart. Trust is essential for the transfer of new technology because producers depend on the agent's word and reputation. By placing the extension agent in a position of enforcing regulations, the agent's ability to gain the confidence and trust of the producers is hampered.

Advantages and Disadvantages

The principal role of extension agents is that of educator. As stated above, whether education is formal or informal the essential purpose of extension is to transfer technology from researchers to producers in whatever manner is most effective. Any other role assumed by an extension agent takes time away from this fundamental activity.

Other roles assumed by an extension agent may result in preventing ancillary business activities from being developed by the private sector. For example, if the extension agents provide a full complement of diagnostic services, at low cost, there would be no market for a private entrepreneur to invest in a water-test kit and begin to develop a business diagnosing water quality problems. If the government, through its extension agents, constructs ponds for producers, provides fingerlings free of charge, or transports feed for livestock, there is no incentive for other individuals to provide these same services. If the government encounters budgetary or political difficulties, the lack of supply of these essential production inputs may endanger the success of the project. On the other hand, unless pond construction is subsidised, the expense of construction may prohibit small-scale farmers from becoming involved in aquaculture.

If an extension agent's defined responsibilities are too broad in scope, it may be impossible for him or her to give good advice in all areas. For example, the background education of most agents is in agriculture or biology, with little or no economics or finance training. But extension agents are often called upon to evaluate credit worthiness of loan applications. The agent may not be adequately prepared to assess credit worthiness properly, and improper recommendations may result.

REFERENCES

Bailey, C., 1988. "The social consequences of tropical shrimp mariculture development". *Ocean and Shoreline Management,* 11:31-44

Ben-Yami, M., 1987. *Aquaculture: the importance of knowing its limitations.* Rome, FAO, Ceres, 112:15-19

Bossard, P., 1987. *Small project review: evaluation of projects.* Kinshasha (Zaïre)

Chakroff, M.S., 1982. *Aquaculture extension services review.* Rome, FAO, Fish.Circ. No. 747

Drewes, E., 1980. *Role of women in small-scale fisheries of the Bay of Bengal. Madras, India, Development of Small-Scale Fisheries in the Bay of Bengal,* FAO/SIDA, (GCP/RAS/040/SWE), BOBP/REP/4, 41 p

Maunder, A.H., 1973. *Agricultural extension: a reference manual* (abridged edition). Rome, FAO

7

Agricultural Extension Reforms

The well-being of the rural population worldwide is invariably linked to the performance of the agricultural sector and to the sector's ability to cope with the challenges that result from rising population pressures, changing demand for food and agricultural products, resource scarcity, climate change, and greater production uncertainty. The *World Development Report 2008* emphasizes agricultural extension as an important development intervention for increasing the growth potential of the agricultural sector in the light of rising demand- and supply-side pressures, and for promoting sustainable, inclusive, and pro-poor agricultural and hence economic development. The call for agricultural extension services is made at a time when the underutilisation of the productivity and growth potential of the agricultural sector for development poses a severe threat for achieving food security and for further reducing (rural) poverty.

The degree of institutional and economic development determines the scope and types of agricultural extension services and the ways in which these services are provided and financed. Ideally, the design of the service provision and funding arrangements reflects the inverse relationship between the stage of economic development and the importance of extension for agricultural development and poverty reduction. The access to well-defined extension services is more important for economies in which agriculture is a major or declining source of economic growth (agriculture-based and transforming economies) but less important in economies in which agriculture is a minor source of economic growth (urbanised economies).

A prominent example for the group of transforming economies is India.The transition from an agriculture-based to a transforming economy was initiated by macroeconomic and nonagricultural reforms in the early 1990s, which triggered unprecedented high growth in nonagricultural (urban) sectors. At the same time, a weak, ineffective, and inefficient extension system and the consequent deficits in knowledge and technology development and dissemination constrained agricultural-sector growth, which in turn caused the share of agriculture in aggregate income to contract from approximately 31 percent in 1993 to about 19 percent in 2003–05.In addition to constraining agricultural sector growth and thus rural development, a weak and ineffective extension system increased the exposure of the agricultural sector to the effects of high population growth, shifts in product demand, natural resource constraints, climate change, and HIV/AIDS, among others.

In order to meet these challenges, India's extension system has experienced major changes since the late 1990s in governance structures, capacity, organisation and management, and advisory methods. The changes involve the decentralisation of extension service provision to the local level, the adoption of pluralistic modes of extension service provision and financing, the use of participatory extension approaches, capacity training of farmers to express their demands, and capacity training of service providers to respond to the demands of farmers, among others. The reform initiatives reflect the view that improvements in agricultural productivity require demand-driven and farmer-accountable, needspecific, purpose-specific, and target-specific extension services.

Birner et al. argue that there is no single optimal or best model for providing needspecific, purpose-specific, and target-specific extension services. The ultimate choice of the agricultural extension approach depends on (a) the policy environment, (b) the capacity of potential service providers, (c) the type of farming systems and the market access of farm households, and (d) the nature of the local communities, including their ability to cooperate. Different agricultural extension approaches can work well for different sets of frame conditions. In order to use extension approaches that best fit a particular situation, the agricultural extension system has to be sufficiently flexible to accommodate the different options. To this end, the recent agricultural-sector reforms have been geared toward creating a demanddriven, broad-based, and holistic agricultural extension system. This has involved the design and introduction of a multitude of integrated measures that, on the demand side, enable service users to voice their needs and hold service providers accountable, and on the supply side, influence the capacity of service providers to respond to the needs of the extension service users.

CONCEPTUAL FRAMEWORK

In order to analyse India's recent agricultural extension reforms, we apply the conceptual framework in Figure 1, first presented in Birner and Palaniswamy. The framework identifies the major governance structures, organisational and managerial characteristics, and frame conditions (for example, socioeconomic characteristics) by which public-sector extension reforms can improve the organisational and managerial performance of service provision, lead to better public-sector governance outcomes, and generate sustainable pro-poor development.

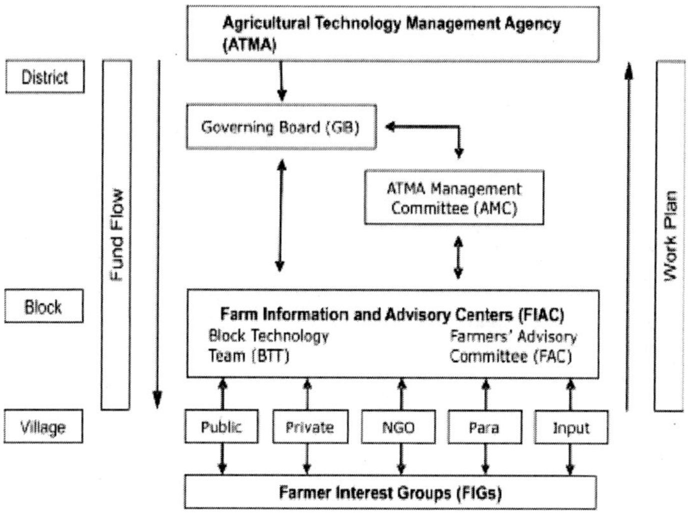

Figure 1. Organisational structure of the Agricultural Technology Management Agency

Public-sector governance outcomes can be evaluated in terms of the efficiency, effectiveness, and long-term sustainability of service provision, regulatory quality, rule of law, the degree of corruption, and equity aspects. Demand-side approaches of public-sector service reforms aim at improving the ability of the private sector (such as farm households and profit-oriented firms) and the third sector(such as nongovernmental organisations, farmers' organisations, and rural women's groups) to demand better governance and to hold public officials accountable by strengthening the voice of clients. To this end, demand-side approaches include policies that increase information and coordination in voting, strengthen the citizens' right to information, and improve the credibility of political promises.

Demand-side approaches of rural service provision also involve policies that promote the political decentralisation of service delivery to local governments, reserve seats in local councils for women, and advocate participatory planning and implementation methods, among others. Figure 1 indicates that demand-side reforms are likely to be more effective if they directly address socioeconomic and sociocultural obstacles that prevent citizens from exercising their voice and demanding accountability.

Strategies to strengthen the demand side of rural service provision will have little effect if they are not accompanied by strategies to increase the capacity of service providers to finance and deliver the respective services, to apply the rules of law and regulation, and to control corruption. Supply-side approaches to public service delivery reforms include the administrative and fiscal decentralisation of service delivery, public expenditure management reforms, training programs for public officials, changes in procurement and audit procedures, and efforts to coordinate the activities of government agencies and departments. Another popular supply-side approach reduces the tasks that are performed by public-sector agencies.

The respective strategies include outsourcing of service provision to organisations in the private and third sectors, public–private partnerships, pluralistic forms of service delivery, devolution of authority to user groups, and privatisation. Recent reform trends emphasize the need for the state to play a coordinating and facilitating role and to create an enabling environment for the private and third sectors. Supply-side approaches also include strategies for cost recovery that aim to improve the financial sustainability of service provision and to strengthen the incentive for clients to demand better services. The usefulness of cost recovery schemes can be debated on equity grounds, especially in absence of specific measures that address this concern. Figure 1 suggests that the success of demand- and supply-side reforms depends on the extent to which reform strategies address the sociocultural characteristics of local communities and the bureaucratic characteristics and incentive structures of public-sector service providers. Because local communities and service providers differ in terms of characteristics, a one-size-fits-all reform approach is an inadequate mechanism for improving rural service provision for rural development. In addition, the structure and scope of reforms also depends on the feasibility of reform implementation. Rather than engaging in ambitious reform programs that address all service delivery problems at the same time, it is often necessary to concentrate initially on those reform elements for which political support can be built. These relationships suggest that reform approaches should center on principles of best fit rather than best practice.

The conceptual framework in Figure 1 points to the role of decentralisation as a governance mechanism to improve the quality of and the access to basic services, infrastructure, and legal and regulatory structures.The importance of decentralisation for good or responsive governance is attributable to the positive effect of decentralisation on the efficient use of resources, with efficiency gains arising from (1) the functional, financial, and administrative autonomy of individual government tiers; (2) role clarity; (3) people's participation; (4) accountability; and (5) transparency, among others. At the same time, decentralisation ensures that each government tier performs those tasks in which it has a comparative advantage. The actions taken by the different government levels are then complementary to each other, with the actions being separated by clear boundaries.

Decentralisation is also subject to weaknesses and is not an accountability mechanism a priori. In fact, economic theory suggests and empirical evidence shows that decentralisation can reduce accountability and increase corruption.Furthermore, decentralisation may lead to welfare losses because of local elite capture and administrative failures. Local elite capture implies that a small share of the population with a disproportionate share of political and economic power resists the changes from decentralisation and participatory policies because of their perceived undue influence on established power relationships.

Local elite capture is associated with inefficient use of resources, inefficient and ineffective targeting of public expenditures and transfer programs, and inefficient delivery of public goods. These aspects amplify inequalities in income distribution, with consequent threats to the economic and political stability of regions. Political and economic tensions may prevail not only within regions but also between them. In fact, they might be stronger if decentralisation reinforces regional asymmetries in income distribution and income growth in the absence of cross-regional income compensation schemes.

AGRICULTURAL EXTENSION REFORM INITIATIVES

In India, major changes in the provision and financing of agricultural extension services materialised in the late 1990s. The reform programs of the late 1990s were motivated by the failure of the World Bankfunded Training and Visit (T&V) model to promote agricultural development by providing effective and efficient extension services in a timely and sustainable manner. Launched in 1977, the T&V model aimed at establishing a close link between agricultural research and agricultural extension. To this end, the model emphasized the role of the state Departments of Agriculture as the instruments through which research institutions should pass on their extension recommendations to the farmers and receive feedback from farmers on the usefulness of new agricultural methods and practices.

The T&V model failed to effectively promote agricultural development mainly because of structural problems in the organisational, financial, and institutional design of the model.Structural problems and the consequent inefficiencies in the delivery of research and extension services resulted from the operation of (1) a hierarchical, classical top-down, one-way communication system and (2) a one-size-fits-all research and extension approach that centered on the institutional, agro-climatic, and socio-economic conditions of irrigated areas but bypassed those of rainfed areas.

The problems inherent in the T&V model were only fully acknowledged in the 1990s. Since then, numerous, mutually reinforcing reform programs have been induced at the central and state levels. These seek to accelerate the development and dissemination of technologies, so as to promote agricultural and rural development through higher productivity growth. The reform initiatives identify and define (1) governance structures and methods for the effective and efficient development (that is, research) and dissemination (extension) of technology and (2) ways for improving the management and organisation of the research and extension system within governance structures (capacity strengthening). At the core of the reform programs are supply- and demand-side elements that embrace principles of decentralisation, transparency, accountability, and e-governance, among others.

Two of the most prominent research and extension reform initiatives were the World Bankfunded 1998–2004 Diversified Agricultural Support Project (DASP) and the 1999–2005 National Agricultural Technology Project (NATP). The DASP initiative aimed (1) to increase agricultural productivity, (2) to promote private-sector development, (3) to improve rural infrastructure, and (4) to increase the income of farmers by supporting intensified and diversified agricultural production and farming systems.

Principal objectives of the NATP initiative were (1) to improve the efficiency of the organisation and management systems of the Indian Council of Agricultural Research (ICAR), (2) to strengthen the effectiveness of research programs and the capacity of scientists to respond to the technological needs of farmers, and (3) to increase the effectiveness and financial sustainability of the technology dissemination system with greater accountability to and participation by farming communities. In contrast to the DASP, the NATP did not emphasize the diversification of agricultural production and of farming systems as instruments to close the productivity gap.

The NATP exclusively focused on research and extension and on an integrated system of extension delivery. Both the DASP and NATP programs involved supply-side and demand-side components of service provision. On the demand side, the DASP and NATP programs aimed at enhancing agricultural productivity, agricultural growth, and rural development by pursuing broad-based, bottom-up participatory developmental approaches that centered on community mobilisation. On the supply side, the DASP and NATP initiatives included (1) public expenditure reforms, (2) changes in planning and decisionmaking processes, and (3) training and capacity building. Some agricultural reform initiatives take explicit actions for improving the access of women to agricultural extension services.

Prominent examples are the ongoing Women/Youth Training Extension Project (WYTEP) in Karnataka, the 1986–2003 Tamil Nadu Women in Agriculture (TANWA) initiative, the 19982003 Training and Extension for Women in Agriculture (TEWA) program in Orissa, and the 1993–2005 Madhya Pradesh Women in Agriculture (MAPWA) project. Implemented under the auspices of the Danish International Development Agency, the projects mainly aimed at strengthening the position of small and marginal farm women in society and at increasing the agricultural productivity and hence income of small and marginal farm women through training in the application of low-cost technologies.Next to the gender-focused projects, reform efforts are also directed toward improving the livelihood of disadvantaged tribal groups.

A case in point is the 1991–98 Andhra Pradesh Tribal Development Project (APTD) of the International Fund for Agricultural Development (IFAD). The project aimed to improve the income and food security situation and living conditions of tribal groups, taking gender, health, educational, environmental, and marketing and credit issues into account. In line with other programs, the IFAD initiative emphasized the importance of (1) productivity-improving farm technology and irrigation systems and (2) participatory planning and implementation as means to improve the livelihood of disadvantaged groups.

In addition to these programs, another prominent reform initiative is the ongoing e-Choupal initiative of the Indian Tobacco Company (ITC). The private-sector program employs information and communication technologies (ICTs) as instruments for improving agricultural extension service provision in terms of outreach (cost) effectiveness, efficiency, transparency, and accountability. In promoting accountability and transparency of service provision, information and communication technologies are designed to facilitate the delivery of demand-side-driven extension services.

Supply-side Reforms

According to the conceptual framework in Figure 1, supply-side reforms emphasize (1) public, private, and third-sector service provision and financing; (2) administrative and fiscal decentralisation; and (3) capacity strengthening and building. This section describes the different supply-side mechanisms for agricultural research and extension, mainly within the framework of the DASP and NATP programs.

The supply-side reforms of the NATP program aim to (1) increase the efficiency and effectiveness of the research system, (2) intensify and support agroecological systems research, (3) improve the efficiency and outreach of technology dissemination, and (4) strengthenhuman capital development and capacity in project management and project implementation.

Closely related, the supply-side reforms of the DASP initiative emphasized (1) technology dissemination, (2) private-sector involvement and public–private partnerships in agribusiness development, (3) rural infrastructure development and marketing support, and (4) project management and capacity building for economic policy analysis. In promoting rural infrastructure development, private-sector involvement and public–private partnerships in agribusiness development, the DASP supply-side reforms were more far-reaching than those of the NATP.

Administrative and fiscal decentralisation

In the case of India, the 73rd Constitutional Amendment transferred fiscal and administrative decisionmaking authority from the central state to the local level. Fiscal and administrative decentralisation are supply-side governance mechanisms to improve local governance in terms of allocative and productive efficiency, service quality and quantity, and transparency and accountability. The DASP and NATP initiative promoted fiscal and administrative decentralisation as mechanisms for improving the effective and efficient dissemination of agricultural technologies and for rationalising and reorienting public extension services.

Under the DASP reform initiative, administrative decentralisation included the transfer of managerial and technical decisionmaking authority from the central to the district level. To this end, DASP operated two implementation mechanisms: the District Project Implementation Committee (DPIC) and the Agricultural Technology Management Agency (ATMA). Both mechanisms were designed to strengthen the co-ordination of the activities of different agencies and to improve the research–extension–farmer linkages.

The DPIC maintained relationships with extension providers and extension users through two channels. First, the DPIC members were representatives of line departments, state agricultural universities, agricultural science centers (*Krishi Vigyan Kendras*), farmer organisations, local governments, and NGOs. Second, the DPIC implemented, guided, and monitored extension activities according to a strategic extension plan that was developed in cooperation with district officers from each line department, technical experts from local zonal research stations, and the trainers from the local agricultural science center. The strategic extension plan defined local research and extension priorities, assessed the relevance and demand for location-specific technologies, and specified the extension activities that were needed to promote farmer-driven agricultural development.

Closely related to the DPIC, the ATMA ensured technology dissemination through fiscal and administrative decentralisation and interdepartmental coordination by bringing together district administration, line departments, NGOs, and local farmer representatives.The stakeholder representatives were directed by a Governing Board, which implemented extension activities according to a strategic extension plan for technology dissemination. In order to increase the outreach of extension services, ATMAs encouraged partnerships with private and third-sector service providers at the district level and below.

Under the DASP, fiscal and administrative decentralisation were also employed as instruments to foster rural infrastructure development and to provide marketing support. The DASP initiative recognised that the development of an intensified and diversified agricultural production and farming system critically depends on a well-developed network of roads and markets. Because rural infrastructure is by and large underdeveloped, the DASP took steps toward promoting the development of rural roads and rural markets. Regarding roads, the reform efforts were directed toward implementing a road development policy that raised funds for the appropriate maintenance and operation of roads.

With respect to markets, policies were designed to reduce the public-sector leverage in the management of rural markets and to improve the operation and maintenance of rural markets by strengthening the responsibilities of traders and farmers. The upgrading of infrastructure also included actions to strengthen the participation of communities in planning, implementing, and maintaining rural infrastructure. In order to increase community ownership and commitment to newly created assets, to improve the operation and maintenance of rural infrastructure, and to increase the representation of traders and farmers, the administrative and fiscal responsibility for rural infrastructure was transferred from line departments like the Department of Agriculture, Animal Husbandry, Dairy Development, and Rural Development to Panchayats. Under the new institutional setup, district Panchayats exercised functional control over district department officers. The process of administrative and fiscal decentralisation was supported by elections to all tiers of Panchayati Raj institutions.

Under the NATP initiative, administrative decentralisation was implemented in the area of technology development (research) and technology dissemination (extension). With respect to technology development, the administrative decentralisation efforts resulted in changes in the expenditure decisionmaking authority and attributed greater financial management responsibilities to the technology development research staff of the ICAR system. The administrative decentralisation of expenditure decisionmaking was driven by the notion that changes in the organisational and managerial structure are needed in order to enhance the efficiency and effectiveness of the research system and hence the longterm growth prospects for agricultural productivity. With regard to technology dissemination, administrative decentralisation was seen as an effective way to support the state Departments of Agriculture in making the technology dissemination system knowledge based and demand driven. The underlying support mechanisms were summarised in the NATP's Innovations in Technology Dissemination (ITD) component, which emphasised the organisation of the technology dissemination system at the national, state, district, block, and village levels.

At the national level, technology dissemination was facilitated by institutions that supported the effective coordination of technology dissemination activities and created a discussion forum on their usefulness (for example, the Technology Dissemination Management Committee [TDMC]) and the Technology Dissemination Unit [TDU]). The

ITD component strengthened the institutions' monitoring, supervision, and evaluation roles for technology dissemination. For instance, the TDU was responsible for the daily management and supervision of the ITD component, while the TDMC supervised and approved the annual work plans of the different project units.

At the state level, the technology dissemination system centered around agricultural line departments. These formed an interdepartmental working group aimed at harmonising project activities among the different line departments at the state level and facilitating and coordinating interactions with the TDMC at the national level. The TDMC included all major stakeholders (representatives of NGOs, women's organisations, and the private sector) and thus provided a platform for these institutions to influence the reform process and the technology dissemination agenda of public institutions. In addition to interaction with the TDMC, the interdepartmental working group also monitored the progress of district-level programs.

Technology dissemination at the district, block, and village levels was institutionalised in the ATMA. The ATMA was an autonomous entity that managed technology dissemination, facilitated decentralisation in planning and implementation, and promoted interdepartmental coordination and demand-driven service provision at the district, block, and village levels by bringing together district administrative entities, line departments, NGOs, and local farmer representatives (Figure 1). Being autonomous, the ATMA had full discretion over its budget and was thus flexible to respond to changing technological and environmental requirements. At the district level, ATMA was supported by a Governing Board and the management committee.

Within its functional framework, the Governing Board identified the programs and procedures for district-level research and extension activities, reviewed the progress and functioning of the ATMA, and approved the Strategic Research and Extension Plan. The Management Committee implemented the district-level research and extension program and conducted participatory rural appraisals to identify problem areas in the implementation of the Strategic Research and Extension Plan. At the block level, programs were implemented through a Farm Information and Advisory Center (FIAC). This institution constituted the operational arm of ATMA and was operated by a Block Technology Team of technical advisors and a Farmer Advisory Committee (FAC). The FAC constituted a platform that encouraged interaction between all key stakeholders, partly by stimulating the formation of commodity-oriented farmers' and women's interest groups at the block and village levels.

Private and third-sector involvement

The extent to which decentralisation can improve the performance of the agricultural research and extension system is likely to depend on the efficiency and effectiveness of public, private, and third-sector service financiers and providers. The relative usefulness of the three service-provider sectors in turn may depend on the degree of excludability and substractability of research and extension activities. Theory predicts that the public sector should only fund and provide the public good components of research and extension, that is, those components with low excludability and low substractability. Similarly, the private and third sectors should only fund services with high excludability and high substractability.

Another supply-side governance mechanism emphasizes the role of the private sector as an instrument to supplement the extension and research efforts of the public sector and to improve the efficiency and effectiveness of service provision. Private-sector participation is seen as a useful means of addressing the research and extension needs of commodity-oriented farmers' and women's interest groups since it utilises private-sector cost advantages, private-sector capacities in research and development, and skills in multi-disciplinary and participatory research. These aspects are perceived to be critical components for improving the accountability of the public sector in agricultural research and extension, for providing demand-driven support services, for encouraging the introduction of improved inputs (such as fertiliser and seed varieties), and for strengthening marketing and postharvest activities and processing.

Both the DASP and NATP initiatives emphasized the importance of private-sector participation in the area of technology development and technology dissemination. The NATP program encouraged private-sector participation within the framework of competitive grants programs.In comparison, the DASP scheme planned to encourage private-sector participation in extension service provision by reducing the role of the government as input supply provider and by facilitating the leasing or sale of government facilities/farms to private-sector firms. In order to strengthen the move to inputs supplied by the private sector and to improve the ability of the market to respond to changing technology needs, state-level government subsidies for inputs and other services were to be phased out, moving to full cost recovery over the life of the project.

The DASP not only supported private-sector participation in input supply and support services in general but took measures to encourage private-sector involvement in agribusiness activities. The aim was to promote the development of intensified and diversified agricultural production and farming systems and to stimulate the vertical integration of smallholder agriculture with input suppliers and agroprocessors. To this end, the program sought to mobilise institutional credit and to facilitate the availability of credit for smallholders' and women's self-help groups. It was thought that the consequent removal or alleviation of credit constraints would encourage incremental investment and the adoption of improved technologies.

In addition to easing credit constraints, the DASP also planned to promote private-sector investment in agribusiness by raising the value of postharvest activities such as agroprocessing. At the core of the initiatives was the so-called Project Development Facility that helped small- and mediumsized investors to establish new ventures or to enlarge current operations so as to exploit market-led opportunities for agribusiness development. The underlying fee-based support services were meant to improve access to technology and technical expertise, to strengthen the linkages between producers and processors, and to improve the availability of information on markets and marketing opportunities.

The DASP and NATP programs also contained provisions for the active participation of thirdsector institutions in technology dissemination. Third-sector participation was to be initiated through contracting-out arrangements. Contracting would delink funding from service provision, and it was seen as an instrument to reduce costs, to improve the cost-effectiveness of public extension services, and to ensure the financial sustainability of extension. NGOs can be important providers of third-sector extension services and useful instruments to generate, refine, and promote need-based agricultural technology for a number of reasons. The involvement of NGOs is important because their work centers on clearly defined and specific objectives. A clear mission and a common objective ensure better teamwork and avoid the uncoordinated implementation of strategies. In addition, a restricted local focus; the consequent closeness to markets; and the adoption of participatory, farmer-centered, bottom-up approaches causes services to be explicitly targeted and demand driven.In addition to contracting out, the DASP and NATP schemes also planned to institutionalise partnerships between public–private and public–third-sector institutions in order to improve technology development, technology dissemination, and agribusiness activities through better access to credit and greater flexibility in program implementation.

In order to enhance the knowledge base and thus the quality of public extension services at the state level, the NATP scheme also advocated partnerships between public institutions. The public-public partnership involved the state Department of Agriculture and the state agricultural university or separate wings of the Department of Agriculture. In Andhra Pradesh, cooperation between the Department of Agriculture and the state agricultural university was enforced through the operation of a District Agricultural Advisory Technology Center in all districts. These centers were supposed to refine technology, make diagnostic visits, and organise field programs in collaboration with the Department of Agriculture and allied departments so as to improve the dissemination of technology, to strengthen research capacities, and to foster human capital accumulation at ICAR institutes. Along a similar line, Punjab Agricultural University fostered adaptive research, training, and consultancy at the district level by having a multidisciplinary extension team in each district, again in collaboration with the state

Department of Agriculture. In Maharashtra, the public–public partnership involved the merger of three separate wings of the Department of Agriculture—the Department of Agriculture, Horticulture, and Soil and Water Conservation—at the operational level. The resulting organisation was deemed to have a broader (multidisciplinary) and more solid knowledge base and a more effective manpower intensity in the field.

Capacity building and strengthening

Reforms of the technology development and dissemination systems also include the rationalisation and reorientation of public extension services through capacity strengthening and building. In the present context, capacity development involves steps toward improving the operational, managerial, and scientific skills and capabilities of individual research and extension workers as well as of the aggregate research and extension system. As will become evident, the existing reform initiatives pursued different approaches to strengthen the capacity of the research and extension system to implement the reform components effectively and efficiently.

The DASP especially emphasized the need for building and strengthening the capacity of those agents who were responsible for project management (line departments) or project implementation (extension workers). Project management was the responsibility of the Project Coordination Unit at the state level and the District Project Coordination Unit at the district level. Project management included the management of project activities, finances, and administration; the provision of technical support to line departments; and the establishment of linkages between the different stakeholders. Operational flexibility and autonomy were provided by the institutional setup of the Project Coordination Unit as a registered society, which offered scope for nongovernmental fund raising.

Project implementation rested with the agricultural researchers and with the extension workers. In order to strengthen the capacities of agricultural researchers for identifying and solving problems, and to enhance the quality, relevance, and accountability of agricultural research, the DASP initiative advocated a competitive grants program. Competitive grants were available for research on key production and processing constraints of the main production systems to state agricultural universities and their zonal research stations, agricultural colleges, ICAR institutes, private-sector institutions, and NGOs. The DASP also contained provisions for improving research coordination. At the state level, this involved institutional changes in terms of governance, staffing, and operational procedures in the Council for Agricultural Research in order to improve policy guidance and to promote and facilitate the cooperation and exchange in agricultural research between institutions.

In addition to strengthening the capacity of program management and program implementation units, the DASP also took steps to improve project monitoring, evaluation, and impact assessment. At the core of the monitoring and evaluation component was the Agricultural Management Center, which had been established under the auspices of the Indian Institute of Management (Lucknow). The Center's main task was the design and implementation of a computerised project monitoring system, which supports the Program Coordination Unit in the management of the DASP program. Within this framework, the Agricultural Management Center (1) prepared economic impact studies, (2) supported the preparation of annual action plans and implementation schedules, (3) identified training needs, and (4) developed strategies for capacity strengthening. Closely related, the DASP program also supported the establishment of an Economic Policy Analysis Unit. The Unit's main objective was to strengthen the ability of the state to develop and analyse the impact of agricultural policies on rural development and to assess the efficiency of public-sector spending on agriculture.

In order to strengthen the capacity of extension workers, the DASP assisted in the redeployment and training of (existing) extension staff (1) to impart new technical skills that help line departments to assume direct responsibility for technology dissemination at the block level and (2) to convey skills in financial and group management that help to effectively operate a demand-driven public, private, and third-sector extension system. The training of extension workers was provided by the state agricultural universities, the State Institute for Rural Development, and the National Institute of Agriculture Extension Management, called MANAGE. These institutions offered professional skill training to extension personnel across all line departments and the agricultural science centers (*Krishi Vigyan Kendras*), forwarded technical information, and supported the technical training of district-level subject matter specialists and of the instructors at the agricultural science centers.

Closely related to the DASP, the NATP sought to improve the intensity, efficiency, and effectiveness of the public agricultural research system by strengthening the capacity of scientists to develop new, productivity-improving technologies that improve the performance of production systems and solve farmers' problems with their farming systems. To this end, the NATP implemented sponsored and competitive grants programs that promoted multidisciplinary and multi-institutional collaborative agroecological research. The multidisciplinary program areas were identified according to a set of national research priorities (improvement of productivity, sustainability of production systems, and precommercial technology development), which were defined to ensure the rational and efficient allocation of resources between different production systems, commodities, and socioeconomic groups. Local research priority areas differed from national research priority areas to the extent that they controlled for local agroclimatic and socioeconomic conditions.

Sponsored research programs were designed to improve the performance of major production systems under intensive irrigated, rain-fed, hill and mountain, coastal, and arid agroecological systems. Thrust areas of research were food security, sustainability, economic growth, equity, and rural welfare, with research centering on natural resource management, postharvest value addition, integrated pest management, and integrated plant nutrient management, among others.

The sponsored research schemes also (1) advocated cross-cutting research on topics that span more than one production or agroecological system and were thus not location specific and (2) promoted strategic (upstream) research on new innovative forms of managing decentralised research activities and research units. Sponsored programs were predominantly implemented by institutions within the ICAR system. In an attempt to strengthen the capacity of domestic researchers and to resolve technical constraints to improved production systems, sponsored programs encouraged the collaboration of ICAR institutes with international research institutes. The sponsored programs were supplemented by competitive grants programs. The competitive grants programs were introduced to promote research on topics that were not covered by the sponsored program and to improve the quality of research and technology development by granting access to financial resources on the basis of competitive bids.

The programs were open to public, private, and third-sector institutions and to public-private and public-third-sector partnerships with the capacity to conduct research in priority areas. The competitive grants programs thus actively encouraged links between the research and development activities of public and nonpublic organisations. Funding was available for location-specific and innovative research, for research on improving the productivity and sustainability of production systems on a time-bound mission mode, for pre-commercial technical development and pilot manufacturing, and for private-sector product and process development.

Similar to the DASP initiative, the NATP program augmented efforts to improve the performance of the research system with efforts to strengthen the capacity of national- and state-level institutions to disseminate, coordinate, monitor, and evaluate technology innovations. Under the NATP, a critical source of human capital development was the National Institute of Agriculture Extension Management (MANAGE). MANAGE's main task was to enable the project-participating states to develop a decentralised, location-specific, farmer-centered, sustainable technology dissemination system. To this end, the NATP program strengthened the role of MANAGE as provider of training in extension management, strategic planning, and participatory rural appraisal to senior managers and trainers at the national and state levels. In first- and second-phase districts, MANAGE also trained the extension workers at the district and block levels.In third- and fourth-phase districts, the extension personnel received training from Agricultural

Management and Extension Training Institutes (SAMETIs). These were established in association with either the state Department of Agriculture or a state agricultural university and had the broad mandate to provide training that supports the implementation of the ATMA model. In addition to the SAMETIs, extension workers also received training from the ATMAs. These agencies offered in-service training and technical backstopping for the extension field staff in order to strengthen the capacity of extension workers to provide broad-based technical advice to farmers. In addition, ATMAs offered training to the staff of line departments in order to improve operational capacities.

Within the framework of capacity strengthening, the reform initiatives also supported the development of information systems. For example, under the NATP, this involved the intensified use of information technology and management information systems and the development of a library information system. The management information system was installed to facilitate project coordination and monitoring, communication, financial management, and performance assessments, among others. The library information system was promoted to strengthen the efficiency and capability of the ICAR research system by improving access to information.

Information and Communication Technology (ICT)

ICT initiatives are driven by the objective to facilitate planning, monitoring, and the effective exchange of information between various agents such as service providers and service users. The advance of ICT programs is attributable to the rapid emergence of extensive telecommunication networks and the rising electrification of even remote areas. ICT programs are perceived to be an essential element in improving service provision and agricultural growth for two interrelated reasons. First, ICT is an effective means for improving governance and service provision by reforming government processes through greater transparency and accountability. Transparency and accountability result from ICT-driven improvements in the access to and the quality of information flows between agents such as service providers and service users. Second, these factors strengthen the scope for participatory actions on the part of service users such as farmers and make technology dissemination demand driven and available to a large number of farm households on a regular basis, mainly through private-sector participation.

The private sector was among the first to rely on ICT as an effective and efficient communication means in agricultural service provision. The most prominent example of a supply-side initiative of the private sector is the e-Choupal initiative of the Indian Tobacco Company (ITC). Reflecting efforts to improve the procurement process for various cropping systems and to reduce the dependence on government-mandated agents (*mandis*), the e-Choupal initiative was launched in 2000. It now operates in 31,000 villages in six Indian states through 5,200 kiosks, extending to 3.5 million farmers.There are plans to up-scale the program to 10 million farmers in 100,000 villages of 15 Indian states by 2010. At the core of the e-Choupal initiative are kiosks in rural farming villages that are equipped with computers with internet access via phone lines or by a Very Small Aperture Terminal (VSAT) connection. As villages often face power supply shortages, each e-Choupal is equipped with a battery-based uninterrupted power supply backup, which provides sufficient energy to run the system twice a day.

Since the system is infrastructure intensive, ITC incurs significant up-front costs in creating and maintaining its own information network in rural India. It also has to identify and train a local (literate) farmer (called a *Sanchalak*) to manage each e-Choupal. Farmers can use the e-Choupal free of charge, while the managing farmer bears part of the operating costs. However, the operating costs need to be compared against the commission that the managing farmer receives for intermediating the e-Choupal transactions between ITC and the farmers. E-Choupal transactions include, for example, accessing the internet for information on scientific farming and best-practice methods, extension, farmer interaction forums, markets, market trends and prices, and weather conditions. Another transaction consists of placing purchase orders for input and consumer goods from ITC or its partners.

The corresponding prices tend to be lower than those available from village traders. In addition, farmers use the e-Choupal to directly sell their crops to ITC. The corresponding prices are higher than the bid prices obtained from government-mandated agents. In paying higher prices to farmers than intermediaries, ITC operates an effective procurement system for crops. At the same time, ITC incurs lower procurement costs, given that the procurement price does not include the commission fees of the intermediaries. Annamalai and Rao argue that the corresponding savings are sufficient to meet the equipment costs from an e-Choupal in the first year of operation.

For reducing the costs of intermediation alone, the ITC e-Choupal system has to be deemed a success. Inspired by the success and popularity of the system, the public sector has tried to increase the outreach of extension and the speed of rural transformation through comparable mechanisms. The NATP initiative, for example, actively encouraged the use of ICT facilities in order to promote the organisational and managerial efficiency of the ICAR research system and in order to strengthen the link between the research and farming system and the research and extension system. The institutional setup involved the establishment of information kiosks and information shops that are operated in collaboration with Farmer Information Advisory Centers at the block level. Closely related to the e-Choupal system, the information kiosks and shops offered direct training of and information for farmers on, for example, crop technology and farmers' rights, loans, and the availability of grants. In order to enable farmers to effectively use the services offered by information kiosks and shops, they received training in the use of ICTs.

Considering the other reform initiatives, ICTs were mainly installed to support the reform projects' efforts in the area of monitoring, evaluation, and capacity building. The NATP, for example, sought to improve ICAR's organisation and management systems by creating an information technologybased library information system and management information system (MIS). The MIS was intended to develop ICAR's Agricultural Project Information System, which aimed to provide a comprehensive overview of agricultural research activities at the national, state, district, and block levels. Under the DASP, information technology was used to develop a computerised accounting and reporting system at the level of the Project Coordination Unit. The respective system advocated a double-entry accounting system, which was interlinked with the related budgeting and financial management system in order to promote the timely generation of summary sheet claims and accounting/fixed assets records.

Demand-side Reforms

A review of the literature shows that the supply-side reforms of the DASP and NATP programs emphasized the need to make technology development and dissemination broad-based in order to promote the productive efficiency of the agricultural and hence rural sectors. As will become evident, the demandside reforms supported this objective. The discussion of the conceptual framework in Figure 1 indicates that demand-side reforms include the empowerment of the rural population by means of political decentralisation, participatory planning and implementation (for example, farmer field schools and the farming system approach), and affirmative action. For the set of reforms that we analyse, participatory planning and implementation appear to be the most prominent modes of empowerment, at least in technology development and technology dissemination.

Demand-side governance structures

On the demand side, the governance mechanisms highlight principles of participatory planning and implementation. Participatory approaches are considered to be useful instruments for increasing the productive efficiency of the agricultural and rural sector by establishing a decentralised, bottom-up, demand-driven, and financially sustainable technology development and dissemination system. By being decentralised, bottom-up, and demand-driven, participatory approaches help (1) to meet region-specific requirements of major production systems and farmers and hence region-specific institutional, agroclimatic, and socioeconomic conditions and (2) to improve the effectiveness and financial sustainability of the technology development and dissemination system by fostering a new division of labor between government departments, the private sector, NGOs, farmers' organisations, farmers' selfhelp groups, and farmer associations, with greater accountability to the farming community.

With regard to technology development under the DASP initiative, the demand-side reforms were driven by the scope and dimension of the supply-side initiatives. The supply-side reforms promoted technology development as an instrument to increase the use of agricultural knowledge and to make new technologies more adaptable and appropriate to farming conditions. In order to control for farming conditions and to account for the needs of farmers, the supply-side reforms of technology development were associated with demand-side initiatives that institutionalised the participation of farmers in problem identification and technology validation. The demand-side governance reforms in technology development were mainly targeted toward location-specific agriculture, livestock, sericulture, and horticulture.

The success of farmers' participation in technology development is dependent on the existence of effective and strong linkages between farmers, research, and extension, which ensured the effective communication of problems as well as solutions. To overcome communication barriers between the different agents, DASP encouraged *Krishi Vigyan Kendras* and extension agencies of the line departments to conduct on-farm research, validation, and demonstration activities. Furthermore, DASP took measures to encourage the participation of farmers in the development of technology through farmers' organisations, farmers' self-help groups, commodity groups, and producers' associations. The group approach will benefit the underlying interest groups if it results in technology recommendations that meet the conditions of a narrowly defined production environment or promotes the more efficient and cost-effective utilisation of resources from economies of scale and scope.

Considering technology development under the NATP initiative, the program promoted technology development via sponsored research on production systems, crosscutting research, strategic (or upstream) research, and competitive grants research. The local priority research themes controlled for local conditions that were identified by using integrated participatory planning approaches, such as participatory rural appraisal schemes at the zonal level. In comparison, the national priority research themes reflected the accumulated experience of technical specialists for production systems research (scientific advisory panels) and for other modes of research (research program committees), rather than the experience of farmers as the end-users of technology developments.

Turning to technology dissemination, the DASP scheme emphasized demand-driven approaches to stimulate the participation of farming communities in the identification of problem areas, in decisionmaking, and in the implementation of proposed interventions. Similar to the reforms in the technology development system, the reforms in the technology dissemination system asked farmers to articulate their demands through farmers' organisations, self-help groups, commodity groups, and producer associations. The driving force behind the mobilisation of farmers were NGOs, which supported the establishment and promotion of farmer interest groups in collaboration with the front line (district, block, and village) extension workers of the line departments and the *Krishi Vigyan Kendras*. To this end, NGOs and the front-line extension workers received training in financial management, group dynamics, group management, participatory concepts, and leadership skills, among others.

While the technology development component of the NATP initiative contained comparatively weak provisions for participatory action, technology dissemination was clearly demand driven and bottom-up. Extension accordingly involved participatory implementation processes at lower tiers of the government. The demand-side of the technology dissemination system was predominantly defined at the block and village levels through the institutional and operational setup of ATMA. Under this agency, programs at the block level were implemented through a Farm Information and Advisory Center (FIAC), which was operated by a Farmer Advisory Committee (FAC).

The FAC hosted all key stakeholders and farmer representatives. This institution's main task was to stimulate the foundation of farmers' groups on the basis of a specific commodity or a general purpose at the block and village level, in order (1) to strengthen the links of farmers to markets, credit, and marketing services through better organisation of farmers and (2) to make technology dissemination and generation farmer driven and farmer accountable. The ultimate objective behind the formation of farmers' organisations, commodity-oriented farmer interest groups, farmers' cooperatives, self-help groups, or women's interest groups was to make farmers and their organisations fully responsible for the technology system.In addition to the FAC, NGOs supported the mobilisation of farmers in voluntary informal interest groups.

Capacity building and strengthening

As opposed to the supply-side reforms, capacity building on the demand side improves the ability of farmers to demand good governance and to hold public officials accountable. In order to be effective, farmers require information on the responsibilities of service providers and on enforcement mechanisms. In addition, the success of demand-side initiatives also depends on the capacity of farmers to identify and communicate technology problems and to implement their solutions.

Under the DASP and NATP initiatives, human capital development of farmers at the block and village levels mainly concerned the effective and efficient use of agricultural input variables (such as seed and fertiliser) and other technologies. The DASP initiative addressed human capital insufficiencies of farmers by organising extensive training programs and demonstrations on the main extension themes of integrated pest management, new varieties and management practices for horticultural crops, animal health, breed conservation, and clean milk production. These activities were provided by line departments and *Krishi Vigyan Kendras.* Like the DASP, the NATP promoted public initiatives toward more intensive training in all project districts of the country through the *Krishi Vigyan Kendras,* zonal research stations, and agricultural technology information centers. The underlying activities included the training of farmers, the dissemination of research findings, and the supply of research products (on a cost recovery basis).

The ability of public-sector training institutions to adequately meet their intended purposes is hampered by (1) tight fiscal budgets and the cutback in available financial resources, (2) the large number of increasingly diversified farmers, and (3) the need to provide training units that reflect up-to-date information. The public-sector training activities were therefore supported by private and third-sector organisations. Third-sector institutions taught farmers to form and manage voluntary informal farmers' organisations and thus created platforms for the effective dissemination of technology innovations. Private-sector organisations provided training and technical recommendations on the efficient and adequate use of inputs to farmers and disseminated farmer-driven and farmer-accountable technologies at the district level. In providing information on the use of inputs and on the scientific management of crops and cultivation practices, among others, the private sector reinforced trends of diversification and thus intensified agricultural activities.

Affirmative action

Affirmative action describes the steps that are taken to empower (rural) people to demand the agricultural and rural services they need, and to make service provision and financing accountable to elected bodies of local government. In India, affirmative action includes specific measures such as training or the reservation of seats in elected bodies of local government for women and disadvantaged groups.

Both the DASP and NATP initiatives contained provisions for the empowerment and mobilisation of women and took actions to promote the participation of women in the planning and implementation process of agricultural research and extension, although gender issues were not a key priority. For example, both programs promoted the formation of farmer and self-help groups for women and required that 30 percent of the farmer representatives on the ATMA Governing Board and the blocklevel FACs are women. The NATP program imposed the additional requirement that 30 percent of the budget for extension and research programs would be allocated to women farmers and women extension functionaries.

Gender empowerment was a key priority of the Danida projects. The WYTEP, TANWA, TEWA, and MAPWA initiatives mainly aimed at strengthening the position in society of women with small and marginal farms and increasing their productivity and hence income by training them in the application of low-cost technologies.

At least during the early phase, all projects implemented the T&V approach. The respective training emphasized the top-down delivery of standardised, predetermined messages. These mainly concerned crop production and were communicated in district agricultural training centers. While WYTEP pursued the T&V approach throughout the project period, the TANWA, TEWA, and MAPWA gradually followed a holistic and participatory farming system approach and widened the scope for agricultural training and extension activities after the initial stage. This included the provision of villagebased training, which is considered to be more flexible to accommodate local-specific agroecological and socioeconomic properties, as it is not institutionalised.

There are differences in the extent to which the TANWA, TEWA, and MAPWA projects broadened the scope for agricultural training and extension activities. For example, TANWA aimed at integrating agriculture with other related activities like crop diversification, horticulture, sericulture, and animal husbandry, while MAPWA only addressed crops. Female extension staff supports technology dissemination at the village level. The training of the women extension workers was mainly accomplished in district agricultural training centers, with a few selected female extension workers also completing a six-month agricultural management and extension course at the Nordic Agricultural Academy in Denmark.

The direct, village-based training built on women, who were trained in technical and communication skills in a planned five-day training course. In order to increase the effectiveness of female extension workers, the projects promoted the integration of the female staff into regular extension services. The integration of the female extension workers into the state Departments of Agriculture was supported by training sessions, workshops, and courses for male staff members. Technology dissemination at the village level was also supported by the groups for farm women and women's self-help groups. Formed with the assistance of female extension workers, the women's groups were meant to create a platform for technology dissemination, shared learning, collective action, and credit and savings activities.

In addition to the gender-focused projects, reform efforts were also directed toward improving the livelihood of disadvantaged tribal groups. A case in point is the 1991–98 Andhra Pradesh Tribal Development Project. The project was implemented in 2,077 villages in four districts that were home to four contiguous and 63,370 tribal households.With annual income levels in the range of Rs 2,660 to Rs 3,770 at the start of the program, the tribal groups faced severe food insecurity. In order to improve the income, food security, and living conditions of tribal groups, the IFAD initiative institutionalised participatory actions of tribal farmers within the framework of self-help groups, cluster-level associations of self-help groups, and village development committees,agricultural consultants, village tribal development associations, and so-called community coordination teams. Village tribal development associations were nodal institutions designed to prioritise the extension needs and to deliver development programs to the community. Community coordination teams were groups of young professionals who were supposed to assist village extension workers in social mobilisation, awareness building, technology dissemination, and the identification of needs, around which development interventions could be built. Mobilisation aimed at strengthening the ability of the tribal population to clearly identify the obstacles to development and to define action plans to help mitigate or even eliminate these hindrances. Self-help groups were also seen as a medium of technology dissemination and as a source of information regarding irrigation, horticulture development, soil conservation, as well as marketing, savings and credit activities.

THE SERVICE DELIVERY EFFECTS OF AGRICULTURAL EXTENSION REFORMS

As systematic program evaluations tend to be unavailable for projects that are implemented by civil-society institutions, the present review of performance assessments predominantly refers to studies of projects funded by the World Bank and Danida. Numerous caveats apply to the underlying studies. First, the existing program evaluation studies mainly highlight the economic effects of the reform initiatives and their program subcomponents but do not identify the impact or the effects of the reform initiatives on the quality (scope and content) and outreach of agricultural extension services. The absence of impact assessment studies can be attributed to the nonexistence of baseline data.Second, the program evaluation studies do not systematically assess the performance of the individual supply- and demand-side components of the reform initiatives. Third, the existing program evaluation studies report nonempirical qualitative or quantitative evidence or both. As the respective evidence does not provide information on the statistical significance of the results, it offers scope for subjective interpretations of the observed relationships. Fourth, the studies tend not to identify the grounds on which the performance of a project is deemed to be satisfactory or unsatisfactory.

Economic Program Performance Assessment

In its implementation completion reports, the World Bank argued that the organisational and institutional changes inherent to the DASP and NATP approach had a positive economic effect. The performance of both programs was judged to be satisfactory given improvements in the prospects for productivity growth.

Under the DASP initiative, the improved prospects for productivity growth were attributed to the use of a farming –system approach, rather than a commodity-specific approach, to technology development and dissemination and to the applied research on user- and location-specific requirements.

These approaches promoted the intensification of the agricultural production system and the diversification into high-value nonbasic farm products. In order to quantify the net economic benefits of program activities, the World Bank implemented a survey that covered both project and nonproject areas and households. The sample from the project (nonproject) area consisted of 333 (164) households from 56 (28) villages in 28 (14) project blocks. Using these data, the net economic benefits of program implementation were summarised in terms of the economic and financial rate of return. The economic rate of return was computed for the overall project and for subcomponents (agriculture, sericulture, livestock, rural roads) thereof. Using information on farm costs and benefits and deducting overhead project costs that are directly attributable to individual components, the economic rate of return was found to be almost equal: 26 percent for agriculture, 25 percent for livestock, and 24 percent for rural roads. The rate of return for the program as a whole was 21 percent. The overall economic rate of return was adjusted to exclude overhead costs that cannot be attributed to individual components. Turning to the financial rate of return, it was computed as the level of farm production and breed improvements after the adoption of new technologies.

The results indicate that (1) the adoption of a representative package of technologies/ practices introduced by the project increased farm-level income by 63 percent per year, and (2) keeping an improved breed increased income from a cow by 125 percent and from a buffalo by approximately 22 percent. The significant increase in income for crops was attributed to four factors: cost saving, productivity gains of approximately 10 percent, diversification into higher-value crops, and increases in cropping intensity from 169 to 203 percent. Unfortunately, the World Bank (2004a) did not investigate the extent to which the rates of return were influenced (1) by the bifurcation of Uttar Pradesh and by changes in world prices for agricultural products, (2) by the lack of leadership in the Project Coordination Unit and in the line departments due to high turnover, and (3) by the lack of reporting and accounting standards during the first years of program implementation and the consequent lack of oversight over the emergence of possible problem areas.

With regard to the NATP program, productivity gains reflected accomplishments in the major objectives of (1) enhancing the organisational and managerial efficiency and effectiveness of the ICAR research system, (2) intensifying and supporting the agroecological research system, (3) improving the efficiency and outreach of technology dissemination, and (4) strengthening the capacity in project management and implementation. The net economic benefits of the NATP program activities were measured in terms of the economic rate of return for the two largest project components: the agroecological research and the Innovations in Technology Dissemination component. Both components accounted for 82 percent of the aggregate project costs.

The production systems research component involved research projects that did not contain adequate provisions for impact assessment. Because of this, the economic and financial rate of return on reforms of the agroecological research system could not be derived in a systematic fashion. Instead, the economic performance was evaluated for three selected major projects: zero tillage, integrated pest management, and household food and nutrition security. The performance assessment employed data from on-farm research for a sample of 690 participating or nonparticipating farmers from 10 states and economic and secondary published information. For these data, the economic rate of return on reforms of the agroecological research system was predicted to be 34 percent.According to the World Bank, the high rate of economic return was driven by the high social returns from the use of researchdriven eco-friendly technologies (such as integrated pest and nutrient management and zero tillage) and their significant impact on the sustainability of land and water use and their negative effect on the use of pesticides and chemical fertilisers, among others.

The net economic effect of the Innovations in Technology Dissemination component of the NATP program was determined by using information from an impact assessment survey that consisted of 1,224 project households and 486 nonproject households. The households held varying sizes of land and were randomly selected from 12 out of 28 districts. To allow for comparisons, project and nonproject households were selected from the same district. Using these data, the monitoring and evaluation report highlighted a 21 percent increase in the level of farm income in districts that operated under the ATMA approach. This increase was assumed to reflect the effects of agricultural diversification and the associated increases in the horticulture cropping area from 12 to 16 percent, in the oilseed cropping area from 3 to 11 percent, and in the herbs and medical cropping area from 1 to 5 percent. At the same time, cereal yields increased by 14 percent, although the area for cereal cropping contracted from 55 to 47 percent during the period 1999–2003. In comparison, districts without ATMA reported an increase in household income of only 5 percent.

Overall, the evidence on the economic and financial rate of return suggests that the DASP and NATP initiatives as a whole succeeded in promoting agricultural and rural development. Unfortunately, the existing program evaluation studies did not identify the relative contribution of individual program components to the overall program response of the agricultural sector. If at all, the evaluation studies only determined the performance of the individual program components.

REFERENCES

Levy, B. 2004. Governance and economic development in Africa: Meeting the challenge of capacity building. In *Building state capacity in Africa*, ed. B. Levy and S. Kpundeh. Washington, D.C. and London: World Bank and Oxford University Press.

Narayan, D., R. Patel, K. Schafft, A. Rademacher, and S. Koch-Schulte. 2000. *Voices of the poor. Can anyone hear us?* Washington, D.C.: Oxford University Press and World Bank.

Nataraju, M. S., M. T. Lakshminarayan, and K. H. Nagaraj. 2001. Role of non-governmental organizations in the new millennium - Case studies. In *Private extension: Indian experiences*, ed. P.C. Shekara. Hyderabad: National Institute of Agricultural Extension Management (MANAGE).

Rajaraman, I. 2000. Fiscal features of local government in India. In *Governance, decentralization, and reform in China, India, and Russia*, ed. J. –J. Dethier. Dordrecht: Kluwer Academic Press.

Reddy, M. N., and B. Swanson. 2006. *Strategy for up-scaling the ATMA model in India.* Annual Conference Proceedings of the Association for International Agricultural and Extension Education 22: 561–569.

Rivera, W., M. K. Qamar, and L. van Crowder. 2001. *Agricultural and rural extension worldwide: Options for institutional reform in the developing countries.* Rome: Food and Agriculture Organization of the United Nations, Extension, Education and Communication Service.

Shankar, K. R. 2001. Private extension in India and a SWOT analysis. In *Private extension in India: Myths, realities, apprehensions and approaches*, ed. P. C. Shekara. Hyderabad: National Institute of Agricultural Extension Management (MANAGE).

Shekara, C. 2001. Private extension: Indian way. In *Private extension: Indian experiences*, ed. P. C. Shekara. Hyderabad: National Institute of Agricultural Extension Management (MANAGE).

8

Agricultural Extension and Food Security

Today, extension is more than it used to be. Its function and tasks are increasingly assumed by multiple public and private organisations. In developed countries, and in countries where extension reform has been pursued, pluralistic involvement of extension providers now exists- including non-profit non-governmental organisations (NGOs), for-profit private companies, rural producer organisations, commercial individuals and associations of extension specialists, as well as national, state and municipal extension services.

A new vision for extension is needed—one that views extension as a main pillar in serving the public good of food security, taking into account the immediate and potential impact of external forces, such as globalisation and trade liberalisation. Future economic and social development needs to be considered in the light of production, marketing and micro-enterprise development of poor people in the rural sector.

NEW VISION FOR EXTENSION

Extension reform requires policy vision and determination, and a nationwide strategy that holds promise of implementation. Whether to decentralise, privatise, or institute contractual arrangements with the private sector, or to promote end-user financing of extension, these are country-specific questions that will require systematic analysis and preparation, and involve gradual change, system coordination and system oversight. Within the framework of such a policy vision, a multimedia communication strategy will also need to be designed and implemented.

MODERN AGENDA

Extension services should be a part of the decentralisation and devolution agenda that engages local government units and grass roots organisations. This facilitates access to broader rural development financing and other rural development and fiscal transfer

programmes, local government financing, and user funding. Decentralisation of the administration of public funds is essential for adopting grassroots approach.

Once the decision is taken to revisit the potential of extension/communication in a major thrust toward a food security initiative, then the next step is to develop a national strategy for extension and rural development information. Let the decentralised extension services be strong in terms of funding and technical competence for implementing the local food security agenda within the context of the national strategy. This responsibility mainly falls on the shoulders of elected local government officials who run decentralised units.

Agendas that Bridge the Divide

Extension information and communication are essential components of the development process; and yet these systems are rarely well integrated into development strategies and programmes. For example, the rush to develop Internet access in developing countries needs to be adjusted to the context of already established and effective communication systems, and attention needs to be paid to bridging the rural digital divide between those who can afford access to the new technologies and those who cannot.

A very important consideration is that information technology tools should facilitate the work of extension agents, not replace them. Stakeholder participation in decision-making processes is crucial and requires multi-sectoral collaboration and partnerships.

Partnership is another way of saying "coordination". There are numerous, and interrelated, actors involved in development. Recognising this pluralistic environment means that extension must become more "extraverted" and cooperate with other agencies within the public sector and with various organisations in the private sector.

Strategy for a New Extension Vision

Strategy involves determining target areas and populations, calculating the potential and capacities of these targets, and developing appropriate programmes—in cooperation with targeted populations—to promote the necessary capacities to meet the needs and demands of the target areas and populations selected. Once these basics are determined, complementary actions need to be undertaken, viz.:

- Design differentiated strategies to reduce poverty and food insecurity.
- Pursue approaches that recognise diverse livelihoods.
- Implement programmes to strengthen producer capacity.
- Focus on development of human and social capital.
- Establish social safety nets to enhance food security of the very poor.

Guidelines for a New Extension Vision

A conglomerate of tasks needs to be undertaken to combat food inse-curity. These tasks will need to be undertaken simultaneously, although each task would necessarily be developed in phases. Several basics appear fundamental to move toward a new vision of extension.

These essentially call on governments to

(1) initiate the proposed policy/strategy process, keeping in view that food security requires long-term planning and commitment rather than ad hoc measures,

(2) consider decentralisation options,

(3) plan system parameters,

(4) promote coordinated actions,

(5) create research/extension teams to respond to community-expressed needs,

(6) review extension/communication involvement in other programmes, such as FAO Special Programme for Food Security,

(7) catalyse and manage programmes and programme activities,

(8) monitor and evaluate,

(9) explore multiple methodologies, and

(10) promote extension communication.

These guidelines are suggestive of how national government can begin to realise a nationwide extension/communication policy and strategy.

Strategies for agricultural and rural development will require situational analyses and needs assessments. Any design to fit the needs and potential of different countries will necessarily need to be differentiated. This differentiation will likely be necessary even within countries and within particular areas and among similar but distinct populations. Thus, hard choices may need to be made at initial stages in determining who will be served, what capabilities will be developed, and what programmes provided. That is why it is important to realise that no single extension methodology is suitable for all situations and for all purposes. The need for developing location-specific

A PLURALISTIC EXTENSION NETWORK

In principle, a pluralistic extension network aims to promote the advancement of "mixed economies" whereby public and private sectors cooperate more closely. A study on the rate of adoption indicates that high rates of adoption of improved agricultural

technologies occur when government organisations, NGOs, and private organisations form partnerships in extending agricultural technologies to farmers.

A pluralistic extension pattern demands that programmes be jointly planned, implemented and evaluated by all service providers, in active collaboration with farmers. The role of the government becomes crucial in a pluralistic extension situation in terms of national policy direction, coordination and quality control to safeguard the interests of farming communities.

A multi-sectoral, integrated extension network will be sustainable only if it proves to be cost-efficient and eventually attains cost recovery. Cost efficiency requires good, transparent management, quality control, and careful monitoring and regular evaluation. Good management requires good leadership and also human resources development that results in strict attention to both the organisation of the partnership with other extension providers and the organisation of public-sector staff and contracted institutions toward technical and group process training. However, training works only when there is motivation to learn as well as to make use of the training received. As with all services, cost is an important issue.

Public financing is critical at first, despite the trend away from subsidies and toward a variety of competitive and contractual mechanisms. In addition to competitive funding and contracting to private service providers, additional sources of funding may be tapped through community development funds, and eventually through user fees and cost sharing. New funding mechanisms are defined in terms of the actors involved, the flow of funds and services, and conditionalities on use of funds. Most of these mechanisms combine empowerment of users with co-financing for specific services, e.g. member fees paid to farmer organisations and levies on agricultural production.

While not immediately realisable with food-insecure populations, these funding mechanisms may be desirable as eventual end goals when and if they promote not only competition between service providers but also contractual relationships between the funding source and the service provider. Funding and accountability may be dramatically altered under these new mechanisms and put farmers in the driver's seat. Community-driven development funds also offer an opportunity for funding extension. Donors and some governments now provide a large share of support to agriculture through these funds.

The process operates as follows: extension micro-projects at the community level are identified through participatory diagnosis involving agricultural staff and approved at that level by a selection committee with a majority of producers' representatives. Producer organisations contract the necessary technical expertise to prepare the micro-projects and implement them with some co-financing from users. In many developing

countries, a mixture of funding and service delivery modalities involving both public and private sector will make sense.

The governments may provide funding while the private companies and NGOs could deliver the extension services. Another modality: a part of the extension system may provide private, fee-based extension advice to commercial farmers and cooperatives while the other part may continue providing free extension advice to subsistence and poor farmers in support of food security—a public good.

FOCUS ON DEVELOPMENT OF HUMAN AND SOCIAL CAPITAL

Food security is a mammoth challenge. The public sector alone cannot finance, let alone deliver, extension services to meet all requirements. Also, as funding has generally been reduced for public sector extension services, field agents have been downsized and those who remain are less able to operate effectively at village level, especially in remote areas.

A reassessment of how to ensure maximum impact from the use of public sector resources is needed. Following assessments and mapping of food insecure areas, a division of labour could be determined whereby different entities undertook distinct efforts either within an area or between areas. Separated agencies, organisations and projects working without coordination will not achieve the goals of the World Food Summit. If food insecurity is to be tackled full-scale, then a concerted integrated national approach is a major first step, one that involves farmer and community at the decision-making level.

Demand-driven extension, i. e. extension programmes based on the needs and demands of food-deficit producers and communities, need to be strengthened through a wider variety of institutional interventions than just public sector extension. If the poor are to benefit from extension, extension reform is needed but importantly, reform that promotes local programmes within the framework of a national integrated food security strategy that helps the poor enter society's mainstream. This is not only a moral and social obligation but also in the economic self-interest of Member States.

Extension is a support and educational agency focusing on changing human behaviour in positive sense, and as such is a very important actor in any national strategy of food security. However, no matter how efficient is an extension system, how qualified and competent its human resources, how generous financing it enjoys and how sound is its operational strategy, extension alone cannot guarantee sustained food security.

While extension is allowed to play its role within the context of new vision, the governments should:

- build platform to promote dialogue and cooperation among relevant institutions and programmes in all sectors to develop a network for food security and income-generation;

- create multi-disciplinary agricultural and rural development teams to respond to rural community-expressed needs;
- build technical and operational capacity of public and private service providers as well as recipients;
- establish and maintain links between policy-makers, support services, small farmers and markets;
- explore multiple programme methodologies;
- create participatory mechanism to upscale proven best practices in food security.

REFERENCES

Ahmed, A. 1982.. *The role of the information system in development.* Studies Series, No. 314. Baghdad, Iraq: Ministry of Culture and Information.

Blackburn, D. J., & Vist, D. L. 1984.. Historical roots and philosophy of extension. In D. J. Blackburn Ed.., *Extension handbook* 1st ed... Guelph, Ontario: University of Guelph.

Chambers, R. 1993.. *Challenging the professions: Frontiers for rural development.* London: IT Publications.

9

Strategic Extension Campaign

A "strategic extension campaign" (SEC) methodology developed by FAO has been introduced in Africa, the Near East, Asia and Latin America. This methodology emphasizes the importance of people's participation (i.e., intended beneficiaries such as fields extension workers and small farmers) in strategic planning, systematic management, and field implementation of agricultural extension and training programmes.

Its extension strategies and messages are specifically developed and tailored based on the results of a participatory problem identification process on the causes or reasons of farmers' non-adoption, or inappropriate practices, of a given recommended agricultural technology or innovation. The SEC technology transfer and application approach is needs-based, demand-driven, and has a problem-solving orientation.

The SEC programme follows a systems-approach which starts with farmers' Knowledge, Attitude and Practice (KAP) survey whose results are used as planning inputs and benchmark/baseline for summative evaluation purposes. In addition, a series of practical and participatory approach workshops are conducted to train extension personnel, subject-matter specialists, trainers and farmer leaders together on the skills of extension programme planning, strategy development, message design and positioning, multi-media materials development, pretesting and production, as well as management planning, implementation, monitoring, and evaluation.

One of the strengths of this approach is in orienting and training relevant extension personnel to apply a systematic, rational, and pragmatic approach to planning, implementing, managing, monitoring and evaluating regular/routine programmes of an agricultural extension service. Empirical evaluation studies (using information recall and impact surveys, focus group interviews and management monitoring surveys) of strategic extension campaign methods applied to specific FAO-supported extension activities conducted, for instance, in Bangladesh and Malaysia (on rat control), Thailand (on pest surveillance system), Malaysia (on weed management), Zambia (on maize production), Malawi, Jamaica, and Morocco (on population education), etc. reported positive changes in farmers' knowledge, attitudes and practices vis-a-vis the recommended technologies as well as significant economic benefits.

This SEC method has been replicated with FAO assistance in many countries in Asia, Africa, the Near East and the Caribbean, with topics ranging from line sowing method of rice cultivation, maize production, cocoa cultivation, tick-borne disease control, contour tillage, population education, ploughing with drought animal power, etc. In addition to various SEC replications within a country, the multiplier effects of its method are felt beyond national boundaries.

USES OF SEC

The Strategic Extension Campaign (SEC) is not an alternative to the conventional extension programme or activity. SEC is, and should be, an integral part of the programmes of an agricultural extension service. The effectiveness and efficiency of such a service could be increased due to SEC's emphasis on its problem-solving orientation, participatory planning approach, intensive extension personnel training, multi-media materials development, and extension manage meet, monitoring and evaluation procedures. Its activities should be carried out by extension personnel and to support the Ministry of Agriculture's policies, strategies and priority programmes. The strategic extension campaign is useful and important to an agricultural extension service due to the following:

1 It Advocates a Participatory Planning Approach
2 It is Needs-Based and Demand-Driven Oriented
3 It Uses Strategic Planning and Integrated Systems Approach
4 It Considers the Human and Behavioural Dimensions
5 It has a Problem-Solving Orientation
6 It Employs a Cost-Effective Multi-Media Approach
7 It Provides Specific Extension Support Materials and Training
8 It has Built-In Process Documentation and Evaluation Procedures
9 Its Method is Applicable to Other Extension Programmes

PARTICIPATORY PLANNING APPROACH

This participatory approach extension method is responsive to intended beneficiaries' agricultural development problems and information needs because its extension objectives, strategies, methods, messages, and multimedia materials are specifically developed based on survey results of their knowledge, attitude and practice (KAP) vis-a-vis the recommended agricultural technologies. Such a participatory approach in planning SEC activities increases the degree of relevance, and thus acceptability, of extension messages or recommendations among intended beneficiaries who are consulted during the planning process regarding their priority concerns and needs. It does not assume the target beneficiaries (i.e. farmers) to be ignorant or requiring all the information there is to know. Rather, it tries to understand and assess farmers' local indigenous knowledge, values and belief systems on farming practices which may be good, need to be improved, or perhaps need to be discouraged. In short, it follows the well-known principles of rural reconstruction: "start with what people already know", and "build on what they already have".

NEEDS BASED AND DEMAND DRIVEN ORIENTED APPROACH

In order to make the best use of available extension resources, SEC activities concentrate on meeting the information, education and training needs of intended target beneficiaries. Rather than providing them with the spectrum of information and skills related to a given recommended technology, SEC activities are geared at narrowing the gaps between knowledge, attitudes, and/or appropriate practice levels of the target beneficiaries vis-a-vis the technology recommendations.

Furthermore, the focus of the SEC activities is to create a demand (through information and motivation approaches) and/or to satisfy the demand (through education and training) among the intended target beneficiaries for the necessary relevant knowledge and skills for adopting the recommended technologies. Such a method needs to apply bottom-up and participatory planning procedures which will give high priority in meeting the interests and needs of the target beneficiaries. Tailoring the SEC messages and activities to the specific needs of the intended beneficiaries would not only increase the chances of success but also would increase the efficiency in resources utilization.

STRATEGIC PLANNING AND INTEGRATED SYSTEMS APPROACH

The SEC method advocates an integrated and holistic approach in extension strategy development, programme planning and management, training, media and/or materials development, and monitoring and evaluation. To ensure its relevance to audience needs, and to utilize its resources efficiently, it relies heavily on both quantitative data and qualitative information obtained from target beneficiaries (i.e., farmers) to assist in problem analysis, objective formulation, strategy development, and management planning.

It applies a strategic planning approach in programming and managing its activities, to achieve maximum outputs or results using minimal inputs or resources in the shortest time possible. SEC activities such as surveys, strategy and management planning, multi-media materials design and development, training, field implementation, monitoring and evaluation are integrated as a system which is also an integrated part of a larger extension programme which has linkages with relevant agencies/units dealing with research, inputs/supplies, training, marketing, etc.

HUMAN AND BEHAVIOURAL DIMENSIONS

In order to minimize heavy "technology-bias" of many extension activities, the SEC method gives adequate considerations to human behavioural aspects, such as socio-psychological, socio-cultural, and socio-economic factors which may facilitate or impede adoption, or continued practice by farmers of recommended technologies. Without sufficient understanding of their positive or negative attitudes and behaviour towards a given technology, the "technology transfer" process would be slow and ineffective, especially if the extension emphasis is on appropriate "technology application" by farmers.

There is considerable evidence to suggest that non-adoption of a recommended agricultural technology or innovation is often related to, or caused by, non-technological factors, such as social, psychological, cultural and economic problems. The SEC method gives due attention to human and its environmental factors which may influence important decision-making process related to agricultural technology adoption and practices. It employs a behavioural science analysis, based on a participatory needs assessments and problem identification of the target audience, in developing appropriate strategies and tactics to overcome or minimize human-related constraints affecting the agricultural technology transfer and application process.

PROBLEM SOLVING ORIENTATION

The SEC is particularly distinguished in that it normally focuses on specific issues related to a given agricultural technology recommendation. Its main aim is to solve or minimize problems which caused non-adoption of such a recommendation by intended target beneficiaries (e.g., farmers). Unlike more conventional extension programmes or activities, it does not "extend" the whole gamut of information on the recommended technology package.

Instead, it selects, prioritizes, and utilizes only the most relevant and necessary information or facts which can maximize the effectiveness of extension efforts to minimize or solve the identified problems of non-adoption of a recommended technology. It stresses on the need to provide strategic, critical and/or "quality" information, which must also include non-technological information, given that the reasons for non-adoption of agricultural technologies are often related to socio-psychological, socio-cultural and socio-economic factors.

Appropriate human behavioural science principles are thus applied to extension problem-solving and in information positioning and utilization, which is responsive rather than prescriptive in nature. The segmentation or classification of extension problems, objectives, strategies and information needs according to a target audience's levels of knowledge, attitude and practice (KAP) in regards to a given recommended technology is not only conceptually important, but practical and useful as well.

Problems related to low knowledge level require different solutions than those related to attitudinal problems. Similarly, strategies for changing negative attitudes on a recommended technology are likely to be different than those for solving incorrect practices in technology application or convincing people to try and practice a recommended technology.

The implications of the different KAP levels would greatly influence the development of problem-solving strategies, message design, selection of multi-media mix, and materials development. Application of a behavioural modification approach using information based on the KAP levels of the target audience for message development, media selection and materials development alone could significantly increase the cost-effectiveness of extension activities.

COST-EFFECTIVE MULTI-MEDIA APPROACH

One of the most common problems or constraints of a national extension service is the shortage of field extension personnel to reach large number of farmers in widely spread geographical areas with inadequate transportation facilities. Moreover, extension workers are usually overburdened with unnecessarily heavy work-load which includes almost everything that has to do with farmers at the village level. Such an over-reliance on extension workers is neither technically sound nor operationally efficient.

Some extension functions for certain purposes such as awareness creation, information delivery, motivational campaigns, etc., can be more effectively and efficiently performed by other means, channels, or non-extension groups, under the coordination and supervision of extension workers. Extension workers' work-load could be reduced by mobilizing appropriate rural and community-based resources, including the increasingly accessible and low-cost mass communication channels to disseminate standardized and packaged extension messages, as well as in utilizing local volunteers to serve as "intermediaries" in reaching farmers.

Such an approach does not imply that extension workers can or will be substituted by these community-resources. Rather, it is a rational approach of using available resources more effectively and efficiently for certain tasks, such as the need to use extension workers for educational or instructional purposes which requires two-way interactions, field demonstrations, group discussion, etc., which cannot be done as effectively by mass communication channels.

This SEC method employs a multi-media approach whereby cost-effective combination of mass, personal and group communication channels and materials are efficiently utilized to reduce extension cost and efforts, and to increase its effectiveness in dealing with a larger number of target audience more rapidly.

SPECIFIC EXTENSION SUPPORT MATERIALS AND TRAINING

Most extension services in developing countries suffer from a lack of relevant and practical extension and training materials to support field activities of their extension workers. Many extension workers rely primarily on their inter-personal communication skills, and thus their time and/or presence during farmers' meetings may not be utilized as effectively and efficiently as it should be. Providing specifically designed and relevant extension/training support materials to extension workers will not only facilitate their tasks and reduce their heavy work-load. It will also ensure a certain degree of quality control in the delivery of technical information or extension message contents.

Experience has also shown that extension workers' motivation, enthusiasm, confidence, and credibility would increase if they are given relevant and attractive multi-media support materials which they could use to improve the effectiveness of their extension and training work. In SEC activities, extension workers are provided with pretested extension and training support materials whose messages are specially designed and developed on the basis of the extension programme's problem-solving oriented strategy plan. Furthermore, these extension workers are also given special training to ensure their understanding of extension strategies, message contents, and management/implementation plan, as well as to when, with whom, and how they should utilize the various multi-media support extension and training materials.

BUILT-IN PROCESS DOCUMENTATION AND EVALUATION PROCEDURES

The advantage of employing a Knowledge, Attitude, and Practice (KAP) survey, as one of the tools for participatory problem identification and information needs assessment, is not only limited to obtaining specific baseline data and inputs for planning extension strategies and improving its management operations. It also provides a benchmark information/data for the purpose of qualitative evaluation, in terms of changes in the levels of KAP over-time. In addition, the SEC activities have built-in evaluation procedures, in the forms of formative evaluation and summative evaluation, for which data/information from the target beneficiaries is essential. It uses various participatory-approach evaluation methodologies including among others, quantitative survey, focus group interview, pretesting, recall test, content analysis, field monitoring, and/or cost-benefit analysis.

APPLICABILITY TO OTHER EXTENSION PROGRAMMES

Most, if not all, of the important principles and techniques employed in planning, implementing, and managing SEC activities are applicable for developing and implementing any extension programme. The SEC's process, operational phases, and implementation steps are essentially similar to that of a regular extension programme. SEC could thus be considered as a "microcosm" of a well-planned agricultural extension programme. It may be safe to assume that if SEC activities can be carried out successfully in a campaign context, which has a very short time period, then SEC processes, methods, and techniques, either partially or holistically, can be incorporated effectively into a regular and institutionalized extension programme which has a longer time span.

Given appropriate training on various SEC principles and techniques, and through direct involvement in undertaking a planned SEC programme, trained extension staff could help in applying a systematic, rational, and pragmatic approach to planning, implementing, managing, monitoring and evaluating regular/routine programmes of an agricultural extension service.

As indicated in a number of SEC evaluation studies, many extension staff who had been trained and implemented SEC activities, have continued to utilize their skills in developing and implementing other institutionalized extension programmes for various agricultural technologies. Some have replicated the complete SEC process, while others applied only certain SEC principles or techniques. These efforts have been appreciated and welcomed by many senior officials of the ministries of agriculture where SEC activities were undertaken, as they could see its concrete outputs, results and impact.

SEC PROCESS AND CONTEXT

The important aspects of the Strategic Extension Campaign (SEC) process, and its conceptual framework as well as the ten operational phases will be described. Some relevant examples on various actual field experiences will be provided to illustrate the suggested SEC operational phases.

SEC Operationally Defined

As commonly known, in the context of agricultural extension, a campaign is one of the methods of extension which can reach a large number of target beneficiaries in a short time period. More specifically, the Strategic Extension Campaign is:

– a strategically planned, problem-solving, and participatory-oriented extension programme, conducted in a relatively short time period, aimed at increasing awareness/knowledge level of an identified target beneficiaries, and altering their attitudes and/or behaviour towards favourable adoption of a given idea or technology, using specifically designed and pretested messages, and cost-effective multi-media materials to support its information, education/training, and communication intervention activities."

In order to complement and improve the programmes of a national agricultural extension service, the Strategic Extension Campaign (SEC) method has given special emphasis on the following aspects:

– A strategic extension campaign is purposive, problem-solving, participatory-oriented, and focuses on a specific issue or recommended technology.

– Its goals are consistent with, and guided by, the overall agricultural development policies and extension programme objectives.

– Campaign objectives are specific and formulated based on intended beneficiaries' felt needs and problems identified through a baseline survey of their Knowledge, Attitude, and Practice (KAP) vis-a-vis the recommended technology.

– A specific campaign strategy is developed with the aim of solving problems that caused non-adoption, and/or inappropriate or discontinued practice, of the recommended technology.

– A strategic planning approach is applied in the process of target audience segmentation, multi-media selection, message/information positioning and design, and extension/training materials packaging, development and production, with a view of obtaining maximum output/impact with the least or minimum efforts, time, and resources.

– Formative evaluation in the form of field pretesting of prototype multi-media campaign materials is conducted before they are mass-produced.

– A comprehensive and detailed campaign management planning is an integral and vital part of the SEC process, and it will not only spell out the implementation procedures and requirements, but such a plan will also be used to develop a management information system, including monitoring and supervision procedures.

– Special briefing and training for all personnel who are involved in SEC activities must be undertaken to ensure that they understand their specific tasks and responsibilities and have the necessary skills and support materials to perform such tasks effectively.

– Process documentation and summative evaluation to assess the progress of implementation and impact of SEC activities are conducted, and whose results are used to improve its on-going performance, and to determine SEC's results and overall effectiveness, as well as to draw lessons learned from such experiences for future replications.

Integrated Process and Systems Approach

The SEC method also advocates the need to carry out extension activities in a systematic, sequential, and process-oriented manner, rather than on an ad-hoc basis. It is a planned extension programme with inter-related activities to be carried out following a management implementation plan by well-trained personnel within a given time schedule. The application of such a systems-approach also points out the need to train staff to master the whole extension process, rather than only some elements of the process or part of the activities.

Staff Training

Without a core-group of extension personnel in a country who are well-trained in specific methods and techniques in planning, implementing and managing SEC activities, it would be very difficult to institutionalize SEC into a national agricultural extension service. Such SEC activities could be initiated and conducted with the assistance of international project experts or consultants. However, the SEC's sustainability cannot be ensured unless a reasonable number of extension staff are specifically trained on all important SEC aspects. They must also have adequate and relevant first-hand experience in planning, implementing and managing such activities.

It is, therefore, imperative that in the process of introducing or replicating the SEC method in a country, training of a national group of extension planners, managers and trainers, subject-matter specialists, communication support staff, and field extension officers on relevant SEC methods and techniques be made as an integral part of the SEC programme planning and implementation.

Due to the inclusion of staff training on SEC process and methodology, significantly higher cost and a longer time would be required for the initial implementation of SEC activities. Such an investment on human resources development would perhaps be one of the most cost-effective inputs which could significantly contribute towards institutionalization of an SEC approach in the quest of improving and strengthening a national agricultural extension system and service.

Planning, Implementation and Management

The Strategic Extension Campaign (SEC) method relies heavily on the participation of all those who are involved in campaign activities, including the campaign staff as well as target beneficiaries, in providing relevant inputs or suggestions throughout the campaign process. By so doing, the SEC activities are more likely to address relevant problems and issues, as well as to provide practical solutions which are based on the needs of target beneficiaries.

It should be noted that not only the target or intended target beneficiaries of a campaign should be consulted and sought for their inputs to ensure the relevance and usefulness of the campaign. Potential campaign implementors, such as concerned field staff, trainers, community leaders, etc. should also be involved and given a role to play in different aspects of campaign planning, implementation and management.

Various means of obtaining their inputs and support can be used, and regardless of the methods, an important purpose for seeking their active participation is to make them "share-holders" of the SEC programme or activities. It is not practical and realistic to expect that all target beneficiaries and campaign staff would be involved in every aspects of SEC activities.

However, through a careful and systematic method of soliciting fairly specific opinions, ideas, suggestions from a sample of concerned groups of people involved in SEC activities, a useful campaign map can be drawn to determine its direction, strategy, types of message contents, training needs, field implementation requirements, etc. A problem-solving and demand-driven strategic extension campaign cannot be properly designed and planned without relevant "feed-forward" information from its intended beneficiaries and concerned campaign personnel.

It is thus imperative that a participatory-oriented method of planning be applied to ensure a needs-based extension campaign which can realistically be implemented and provide benefits to its intended clientele.

Strategic Planning of Extension Campaigns

The effectiveness of an extension campaign depends very much on its strategy which should be specific, systematic, and well-planned. In developing a campaign plan, the use of a conceptual framework which can guide the planning process in a systematic, rational, and strategic manner should be considered. A good conceptual framework can help in the application of relevant theories, principles or research findings in the planning and development of sound extension strategies and plans.

Planning is defined as a process of identifying or defining problems, formulating objectives or goals, thinking of ways to accomplish goals and measuring progress towards goal achievements. A good extension campaign plan has a strategy which reflects the target beneficiaries' identified problems/needs and the way information, education (i.e., training) and communication will be used in solving such problems or meeting the needs. Such a plan must outline the management actions to be taken in implementing the strategy. Thus, in this context, campaign planning has to include both strategy planning and management planning.

Strategic planning can be operationally defined simply as the best possible use of available and/or limited resources (i.e., time, funds, and staff) to achieve the greatest returns or pay-off. Strategic planning is also an approach to anticipatory planning in order to reduce or overcome some uncertainties in decision-making process by prioritizing actions or interventions which may produce the most likely positive outcomes or results. A strategic extension campaign plan should provide specific guidelines and directions in making information, education and communication activities operational. It must be constantly reviewed, especially at the implementation stage.

Adaptation or modification of the plan may be required because of specific local conditions/problems or alteration of the policies/objectives which guided the original plan. The plan should thus be flexible and ready for necessary modification as suggested by feedback results/information (e.g. through process and/or formative evaluation, including pretesting) in order to improve the strategy or the management of campaign implementation activities.

The process of developing a strategic extension plan can be divided into two major parts. The first part is the process of strategy development planning. The second part is the process of management planning. To provide a systematic approach in developing a strategic extension campaign plan, a generic conceptual framework is suggested based on a 10-phase circular model. The suggested process of developing a strategic extension campaign plan is described below, adapted from the of conceptual framework originally proposed by Adhikarya:

- *Part I: Campaign Strategy Development Planning*
  - Phase 1: Technology and problems identification and information needs assessment
  - Phase 2: Campaign objectives formulation
  - Phase 3: Strategy development and information positioning
  - Phase 4: Audience analysis and segmentation
  - Phase 5: Multi-media selection
  - Phase 6: Message design, development, pretesting and materials production

During a campaign strategy development planning, as much formative evaluation as possible should be included as a built-in component in all the above-mentioned phases, especially in phases 4 to 6. Formative evaluation in this context means the process of testing the suitability, appropriateness or effectiveness of campaign strategy and plan, including its multi-media messages and support materials, preferably before full implementation, in order to ensure good campaign performance or results.

When a plan for a campaign strategy is completed, it must be translated into action. At that stage, the task of a extension campaign planner shifts from strategy development to management planning. To transform extension campaign strategies into extension campaign activities, management objectives must be identified clearly to include at least the following elements: what the action is; who is to carry out the action; how the action is to be carried out; how much resources will be needed and how to obtain such resources; when the action is to be accomplished. In addition, management objectives should set a standard for measuring progress and impact of implementation. Thus, the following phases are necessary in campaign management planning:

Part II: Campaign Management Planning

- Phase 7: Management planning
- Phase 8: Training of personnel
- Phase 9: Field implementation
- Phase 10: Process documentation and summative evaluation

These four phases of campaign management planning should be supported by a management information system to provide planners with regular and up-to-date information for at least the basic components of management objective: who will do what and when. There are three kinds of management activities for which such information is needed to make effective decisions: personnel, finance and logistics. It should be noted that management information system is useful so long as it does not create an unnecessary burden on extension and training staff; distracting them from their basic information, motivation, and education tasks.

Phase-1: Technology and problems identification and information needs assessment

The planning of a comprehensive extension campaign strategy requires an adequate and accurate set of baseline data including information on recommended technology(ies), identified problems, and information needs. Many types of baseline data need to be collected and analysed. Without such information, a feasible, efficient and cost-effective extension campaign cannot be planned and developed properly. In particular, such information is critical for formulating campaign objectives, developing strategies, selecting and prioritizing extension/educational contents, positioning and designing messages, and evaluating the effectiveness of campaign activities.

Extension planners and practitioners are often forced to make programming or management decisions guided mainly by assumptions or personal experience which might be invalid, and could waste programme resources. Such counter-productive decisions may stem from inadequate planning and/or lack of baseline data and other relevant "feed-forward" information (i.e., data about target beneficiaries' problems, needs or interests on which campaign messages should focus).

Phase 2: Campaign objectives formulation

In order to be effective, strategic extension campaign should be an integral part of a given agricultural extension system or programme and its function should be to support such extension activities. The specific campaign objectives should reflect the extension system or programme goals, respond to the needs of the programme and its target audience and help in solving the problems encountered in achieving such goals. It should be pointed out, however, that strategic extension campaign objectives are not necessarily the same as the agricultural extension system or programme goals which are expected to be the ultimate results of the whole extension undertaking. The achievement of the campaign objectives is a necessary, but not a sufficient, condition for achieving extension system or programme goals.

An extension programme goal should be explicit in specifying what is to be accomplished. For instance, the following two inadequate goal statements give only the general or operational elements to be achieved:

- To provide irrigation for rural people".
- To drill 4,000 ring wells and 2,000 tube wells by August 1994".

Those descriptions of goals can be made more comprehensive and specific, as well as made to reflect the actual scope of the programme, as follows:

- To increase the number of small farmers in districts X, Y and Z using water from the wells to irrigate their farmland from the present 100,000 to 175,000 small farmers within two years".

With such a goal, the extension programme could not be considered successful if it had only drilled 4,000 new ring wells and 2,000 new tube wells within two years although such an operation or activity was important and essential. It might well be that the water from the wells could not be used for irrigating farmers' land but, perhaps only for household purposes, or not at all. Even when the water might serve irrigation purposes, perhaps it would be mostly the large farmers who were using it rather than small farmers.

The ultimate goal of the above programme or project is not only to drill new wells but to persuade small farmers to utilize water from the wells in irrigating their farmland. Since a campaign activity is an important support component of the extension programme, the chances for success in meeting the programme goal are increased if the campaign objectives of the programme or project are accomplished. It is thus also useful and important to distinguish extension campaign objectives from the broader extension programme objectives.

Strategic extension campaign objectives are usually very specific and aimed at increasing knowledge, influencing attitudes, and/or changing practices of intended beneficiaries with regard to a particular action recommendation or technology. Campaign objectives should specify some important elements or characteristics of the extension activities which could help to provide a clear operational direction, and facilitate a meaningful evaluation of the campaign. Some of those elements are, at least, the following:

1. the target beneficiaries;
2. the outcome or behaviour to be observed or measured;
3. the type and amount/percentage of change from a certain baseline figure expected from the target beneficiaries;
4. the time-frame;
5. the location of the target beneficiaries.

It should also be pointed out that if the time frame and/or location of target beneficiaries are not explicitly mentioned in a extension campaign objective, it is usually understood that they follow the time-frame or duration and the target location/area of the campaign itself. In the case of the above-mentioned irrigation programme, examples of strategic extension campaign objectives which would help in achieving general extension programme goals could include the following:

a) to inform within one year at least 65 percent of the small farmers in X, Y and Z districts about the procedures and benefits of an irrigation system using ring and tube wells;

b) to reduce the proportion of small farmers in districts X, Y and Z who have misunderstandings and misconceptions about the cost and technical requirements of drilling and building ring or tube wells, from the present 54 to 20 percent in one year;

c) to increase the proportion of small farmers in districts X, Y and Z regarding their positive attitude towards the practical and simple use of the irrigation system to water their farmland, from the present 32 to 50 percent within two years.

d) to persuade small farmers in districts X, Y and Z to use water from the wells to irrigate their farmland, and to increase the level of such a practice by them from the present 20 to 35 percent in two years.

It should be noted that campaign objective (d) above is very similar to the stated extension programme goal. While this objective may be considered and included as one of the campaign objectives, it should be realised that the accomplishment of this objective is not entirely dependent on effective campaign strategies or activities. There are other "non-extension" factors which might influence the accomplishment of this objective, such as availability of inputs or funds, type of seeds utilized and land conditions. A campaign is only one of the many factors which can contribute to the successful achievement of a programme or project.

Phase 3: Strategy development and information positioning

A systematic and strategic campaign plan is essential to the efficient achievement of campaign objectives. The first step in a strategic extension campaign (SEC) planning exercise is to identify clearly the problems which may impede or obstruct adoption of the suggested idea, innovation or technology. Data from baseline/KAP surveys, including problem identification and needs assessments results should be analysed carefully. The central issue(s) or problem(s) which might impede progress in achieving the extension goal should be identified.

Phase 4: Audience analysis and segmentation

One of the most important elements in a campaign is the target audience or beneficiaries: who they are, where they are located, why they are chosen as target beneficiaries and what information contents or messages should be communicated to them. Analysis of a target audience is an integral part of designing and planning a campaign strategy. In such an analysis, certain types of information or data are needed; for instance, the size and location of the target audience, their socio-economic profile and their socio-cultural profile.

Characteristics, interests and information needs of the target audience might be different, so audience segmentation into several different target groupings is usually necessary. For each target group, a specific campaign strategy may be required. In this phase, it is also important to prioritize the target beneficiaries as to which group(s) should be reached first or be given the most intensive campaign treatment.

Phase 5: Multi-media selection

Many campaign experiences and empirical research studies on communication media effects have shown that the use of multiple media channels which include a combination of mass, group and interpersonal communication, if appropriately selected and utilized, is usually more cost-effective than the use of a single communication medium.

An important aspect in employing a multi-media approach is the proper selection of available channels in order to avoid redundant or superfluous media usage and to optimize the level of multi-media support required. Thus, a multi-media approach does not mean that all available communication channels should be utilized. For instance, the degree of emphasis in utilizing mass, interpersonal or group communication channels depends on target beneficiaries' KAP levels and campaign strategy priorities.

In developing an appropriate multimedia mix, results of audience analysis should be considered, especially on information-seeking habits, preferred information sources, media access or ownership, media consumption or usage patterns, communication network interactions and group communication behaviour. The rationale behind the use of a multi-media approach is that a coherent, coordinated, and reinforcing system of communication should be able to address specific but varied information problems and needs of target beneficiaries.

In addition, since there is no information medium which is effective for all communication purposes or for all types of target beneficiaries, a multi-media approach is considered as a viable alternative. Another reason for employing a multimedia approach is the need to make the extension system more efficient given the various information, educational and communication objectives of a campaign.

Phase 6: Message design, development, pretesting and materials production

The effectiveness of a campaign, to a large extent, depends on the relevance, validity and practicality of the information or messages communicated to the target beneficiaries. Sometimes, even though the main content of the information is useful and technically sound, the campaign information may not be well received or understood by the target beneficiaries if it is not presented properly to them.

Campaign messages must be able to attract the attention of the target audience, be easily and clearly understood, and be accurately perceived by them. In general, it is often assumed that the degree of message effectiveness is a function of the amount of reward that the message offers and the level of efforts required by the audience to interpret, perceive and understand the message. In message design and development processes, given the formula:

$$\text{M.e. (Message effectiveness} = \frac{\text{R(reward)}}{\text{E(efforts)}}$$

then the option that is often more within the control of extension campaign planners is to decrease the level of efforts required by the target audience in interpreting, perceiving and understanding the campaign messages.

Another factor which may influence the effectiveness of a campaign message delivery is the competing messages that try to reach the same target audience. Rural population in many developing countries are "target beneficiaries" of numerous social, economic and political development programmes which tend to overburden them with various types of information. To make matters worse, the messages often conflict with one another, thus further confusing target beneficiaries.

For instance, a nutrition campaign message may suggest "establish water/fish ponds" while a health campaign message may recommend "get rid of stagnant water". Given the "information overload" situation, which has become increasingly common among rural population in many developing countries, a campaign message needs to be strategically "positioned" in order to "stand out" in the crowd.

Otherwise a message may go unnoticed and has no impact even though it is technically useful and relevant for the target audience. Positioning a campaign message strategically and effectively into target beneficiaries' minds requires a well-planned, creative and innovative approach of presenting the message. First of all, the relevance and validity of the message must be ensured.

Once a theme has been identified, messages need to be developed and treated further for effective "packaging", utilizing various social and psychological appeals, in order to make the theme attractive and persuasive. Depending on the campaign objectives and strategy, the following examples of the more commonly used social and psychological appeals can be applied in packaging campaign messages:

- fear-arousal appeals
- incentive/rewards appeals
- testimonial appeals
- community or group/peer pressure appeals
- role-model appeals etc.
- authoritative appeals
- emotional appeals
- civic-duty appeals
- morale-boosting appeals

- common-man appeals
- guilt-feeling appeals

Messages which have been packaged with appropriate appeals can be given different treatments according to the needs, objectives, and strategies of a campaign. For instance, a message presentation can be given one or more of the following treatments:

- serious/formal
- humorous
- popular/informal
- one-sided (i.e. pros or cons only)
- two-sided (i.e. pros and cons)
- repetitive
- aggressive/confrontational
- direct
- indirect
- conclusion-drawing
- fact-giving etc.

The packaging of messages using the above types of presentation appeals and treatments should also take advantage of, or capitalized on, the particular strength or positive attributes of the intended delivery medium. For instance, the advantages of radio differ from those of a leaflet or a filmstrip, and so on. Each medium has its own features which can be utilized to different effects through appropriate message design with the appropriate appeals and treatments. In general, message appeals using fear-arousal or emotional approach, for example, may be best expressed by a visual medium, whereas fact-giving or conclusion-drawing types of message presentation treatment are more likely to be effective if presented through a printed medium.

Therefore, in designing and developing messages, the intended medium to be utilized to deliver the message must be also be taken into consideration. In order to ensure that target beneficiaries correctly interpret, perceive and understand the meaning of a message, a pretesting exercise would be very useful and should be undertaken. Such an exercise is needed especially if visual materials are intended for an illiterate or lowly-educated audience (i.e. less than three years of formal schooling) who might also be illiterate visually because symbolic representation must be learned and experienced in order to be understood.

A message which has been developed and packaged on a prototype basis should be pretested before final production with a small sample of the actual target beneficiaries. A message pretesting exercise can be done in a very short time with limited cost and simple methodology, and may result in a significant improvement in the effectiveness of the message as well as a considerable saving of resources.

Phase 7: Management planning

A good campaign strategy plan does not automatically result in an effective campaign implementation. Unless a management plan is drawn up to specify how the strategy should be put into operation, a campaign is likely to be unsuccessful because the required logistical arrangements and support materials may not have been made or completed on time or according to the specifications of the strategy plan.

There are at least two major elements with which a management plan should be concerned. The first element deals with the assignment of responsibility regarding the specific tasks to be undertaken to conduct the field implementation of the strategic campaign plan.

Another element of a management plan deals with the organization and coordination of planned multi-media activities as well as the proposed system of campaign materials distribution and utilization. Guidelines on the specific recipients and locations, as well as the utilization procedures of such multi-media support materials should be provided in the management plan. These guidelines and procedures are not only needed to facilitate campaign field implementation, but also needed for management monitoring purposes.

One of the main objectives of management planning is to provide campaign organizers with a systematic and comprehensive procedure for mobilizing available resources (personnel, finances and time) effectively and efficiently according to the campaign strategy plan. Since campaign implementation depends to a large extent on persons who are mobilized to carry out the different aspects of the campaign activities, a plan for managing training activities according to the specific campaign training needs is also required.

The implementation of a multi-media extension campaign can only follow the planned strategy if: (a) the media materials are developed and produced as planned; (b) the combination of campaign media are mobilized and coordinated as suggested; and (c) the campaign support personnel are trained accordingly.

A management information system which monitors the campaign programme implementation can help provide campaign organizers with regular and up-to-date information for improving campaign management and performance. But the overall assessment on the impact or effects of campaign implementation has to be determined by a specific, and preferably built-in, campaign programme or summative evaluation.

Phase 8: Training of personnel

While the need for training of campaign personnel is obvious, past campaign experiences have shown that in many instances training has been neglected. The implementation of most campaign activities requires new, different or more specific tasks and responsibilities to be added to the routine workload of the persons who have been mobilized for a campaign. However, in previous campaign activities, it has often been found that a considerable number of the campaign workers were not well-prepared for the specific tasks they were expected to perform. In some cases, they were not even adequately informed about the goals and specific issues of the campaign because their involvement with campaign activities was often conveyed through an administrative order. Such a bureaucratic approach to mobilizing campaign personnel, especially the front-line campaign workers, appears to be an inadequate means for ensuring that campaign workers are motivated and functionally capable of performing the additional tasks required of them to support the planned campaign activities.

Training sessions, or at least orientation meetings, for the various types and needs of campaign workers should be specifically planned and conducted before implementing a campaign. Such training activities are especially important if new or different skills and knowledge are required of campaign workers in performing their tasks effectively. Campaign workers should also be briefed adequately on how to integrate the campaign-related tasks with their routine work programme or activities, in addition to the purpose and value of their participation in the campaign programme.

Phase 9: Field implementation

The most important element in ensuring that a campaign is implemented as planned is the appropriate monitoring and supervision of campaign workers' performance and the extension campaign delivery system. Such a task can be facilitated by a good management information system which is able to provide campaign organizers with rapid feedback on various important campaign activities and thus can help in readjusting or changing campaign strategies if considered necessary.

Another important task in campaign field implementation is the proper coordination of various activities which sometimes need to be carried out simultaneously. Coordination linkages must be carefully developed, especially if several agencies are involved in executing different aspects of the campaign activities. One of the most frequent problems found in campaign filed implementation is the untimely delivery, and often unavailability, of inputs or services required for the adoption of recommended technologies or action by the target beneficiaries who have been motivated and persuaded by the campaign. Such a problem may lead to a "frustration of rising expectations" among some members of the motivated target beneficiaries which in turn might undermine the credibility of the campaign recommendations and campaign workers.

Effective programme implementation also requires proper execution of activities within the estimated time period. A delay in one of the usually interdependent multi-media activities of a campaign may have chain-reaction effects. In planning the implementation of campaign activities, a realistic time estimate for the completion of an activity required to support the campaign should therefore be considered.

Phase 10: Process documentation and summative evaluation

Many people who are involved in rural development programmes have increasingly realized the need for, and the importance of conducting a summative evaluation. The main purpose of such an evaluation is to assess the performance, effects and impact of a campaign. Unlike formative evaluation which is normally conducted at the planning stage or early stages of campaign implementation, a summative evaluation is almost always conducted near or after the conclusion of a campaign. Whereas formative evaluation findings are often utilized to improve campaign strategy or performance during its implementation, the results of summative evaluation are normally used to determine whether the campaign has accomplished it objectives and if an improved or expanded campaign should be undertaken as a follow-up programme.

To ensure that a summative evaluation is conducted properly and that its results are relevant to the campaign objectives, summative evaluation activities should be considered as a built-in component and an integral part of the campaign process. The findings of such a summative evaluation should be used as inputs to formulate new or improved campaign objectives or to help set up new baselines or benchmarks for future campaigns of a similar nature.

It is also useful to conduct a process documentation which points out critical issues and decision-making requirements in undertaking the SEC activities. Through a chronological description and analyses of successful or less-successful decision-making process made during planning, implementation, and management of the campaign, important lessons can be learned, and technical and management operation generalizations can be suggested, for future replications and expansion of similar activities.

Regardless of the outcome of a process documentation and summative evaluation exercise, the lessons learned (both positive and negative aspects) are valuable in designing or planning future campaigns which have similar objectives. Therefore, efforts to disseminate and share summative evaluation results of a strategic extension campaign (SEC) among extension planners, managers, and trainers should be encouraged and pursued. The publication of this book, for instance, is an attempt to share the experiences and results of SEC activities.

REFERENCES

Adhikarya, Ronny 1978.. "Guideline Proposal for a Communication Support Component in Transmigration Projects". A consultancy report prepared for FAO/United Nations project 6/INS/01/T., Rome, Italy: FAO.

Adhikarya, Ronny and Everett M. Rogers 1978.. "Communication and Inequitable Development: Narrowing the Socio-Economic Benefits Gap". In *Media Asia,* Vol. 5, No. 1.

Adhikarya, Ronny and John Middleton 1979.. *Communication Planning at the Institutional Level: A Selected Annotated Bibliography.* Honolulu, Hawaii, USA: East-West Communication Institute, the East-West Center.

Boonlue, Thanavadee 1987.. "Knowledge, Attitude, and Practice on Pest Surveillance System in Chainat Province, Thailand".Bangkok, Thailand: Chulalongkorn, University, October 1987.

Boonruang, Pote and P. Chunsakorn 1988.. "Management Monitoring Survey of the Pest Surveillance Campaign on Rice".Bangkaen, Thailand: Katsetsart University, June 1988.

FAO 1990.. *Report of the Global Consultation on Agricultural Extension.*A proceeding of an international expert meeting held in Rome, 4-8 December 1989. Rome, Italy: Food and Agriculture Organization FAO. of the United Nations

Mohamed, Ramli 1989., "Strategic Extension Campaign Experiences on the Rat Control Campaign in Penang". Penang, Malaysia: Universiti Sains Malaysia Science University of Malaysia., April 1989.

10

Role of Women in Agricultural Extension

The crucial importance of women's contribution to food security in developing countries is widely recognised. In most developing countries, rural women are the mainstay of small-scale agriculture, the farm labour force and day-to-day family subsistence. Yet rural women are faced with a number of constraints. They have more difficulties than men in gaining access to land, credit and extension services.

Development interventions to improve the economic roles of women have also so far had limited success. Up recently, the overriding concern of such interventions for rural women remained one of welfare and home economics programmes, mainly through women-specific projects or women's components in multi-purpose projects.

In many cases, however, development projects have not taken adequate account of women's responsibilities, participation and priorities in their specific local conditions, constraining the achievement of the objectives of the programmes, or leading to negative effects on women and families. In recent years there has been an increasing recognition of the need to integrate women into mainstream development efforts.

The economic rationale behind this approach is that the full use of productive potential of human resources cannot be realised if women, who make substantial contributions to food output, do not have adequate access to resources, productivity enhancing inputs and services. This approach focuses on gender differences primarily in division of labour in production, or income-earning activities, and access to and control over resources and assets.

Gender analysis also involves the desegregation of women's roles and responsibilities by socioeconomic class, agro-ecological environment and farming system, culture and ethnic group and, for each of these categories, by age and marital status. This shift to a gender approach has been primarily due to the overriding attention of governments to the need for socioeconomic policies to take into account the role of and the impact on women.

Many governments are increasingly recognising, for instance, that policies such as price incentives cannot be fully successful in stimulating agricultural production if the institutional arrangements prevent women producers from getting the benefits. Almost everywhere, women are responsible for processing, storing and preparing family food. Rural women also fetch water and firewood for the family.

Above all, women are engaged as family labour in agriculture as well as wage labour and in other income-earning activities, and generate a substantial proportion and sometimes even all of the basic daily food for the family. They are usually responsible for small livestock, often for small ruminants and sometimes for large animals that are not on free range.

In many countries women play an important role in fishing in shallow waters and in coastal lagoons. It is also observed from many studies that, although there is a wide diversity in household production patterns, women in all regions play a predominant role in household food security through agricultural and food production. It is estimated that a significant proportion of women in developing countries spend up to two-thirds of their time in traditional agriculture and marketing. Their working hours tend to exceed those of men.

Women in rural areas grow at least 50 % of the world's food. They work in all aspects of cultivation, including planting, thinning, weeding, applying fertiliser and harvesting. Women are also active in post-harvest activities and in livestock production. In Africa, it is estimated that women represent from 30% to 80% of the agricultural labour depending on area and socio-economic class.

In most parts of this region, women have been traditionally responsible for food crops and men for non-food cash crops. This division of labour between food crops and cash crops, however, is never clear cut. Often women help their men-folk in cash crop production. The contribution of women to agricultural production and household food security in the Near East varies widely from country to country, but in many of the countries their contribution is substantial.

Where permanent drought has forced men to migrate, women's participation in agricultural tasks that were traditionally done by men has increased. In some countries women perform laborious and repetitive tasks all the year round, whereas men work during ploughing and threshing periods only.

In several other countries, irrigation, pest control and to a large extent fertiliser application are the exclusive tasks of men, while threshing and marketing are the exclusive tasks of women. In the region as a whole women are heavily involved in livestock, particularly small animal and dairy production. In Asia, in nearly all rice-growing areas men traditionally perform such activities as preparation, ploughing, irrigation and leveling of the fields. However, sowing, transplanting and weeding are usually women's work. Harvesting and threshing and transporting of grain from fields to home is done by both men and women, while drying, cleaning and processing of the rice are done by women.

A number of cases show that the contribution of women to agricultural production in terms of number of operations performed and number of hours worked is in most cases greater than that of men. Women in Latin America also are heavily engaged in food production, but to a lesser degree than in Africa and Asia. Surveys in parts of Colombia and Peru show that female participation rates in agricultural field tasks range from 25% to 45%.

While in the more commercialised agricultural regions almost all households have at least one woman who participates in field work, female participation in the regions characterised by traditional agriculture is more in processing activities than in direct field work. Women in these regions have, however, a high participation rate in animal husbandry, which in some areas contributes as much as 30% of the household income.

Women form a large segment of the agricultural work force. As such they deserve increased attention of agricultural extension services in every developing nation. There is a need for an action-oriented plan to reach the millions of women in agriculture who fill the bread baskets of the third world and contribute to their exports.

WOMEN AND AGRICULTURAL LABOUR

The 1983 COAG paper contains a great deal of data which clearly show that nearly half of the officially enumerated agricultural labour force in many developing countries are women. Studies also show that a significant proportion of women spend up to two-thirds of their time in traditional agriculture and in marketing. Their working hours tend to exceed those of men. In some countries women are the principal labour force on small farms and perform the major share of hoeing weeding transporting storing processing and marketing. Their share in cash crop farming is also significant. Yet agricultural extension networks often by-pass women producers. This means for women that they are disadvantaged and implies for governments that resources made available for extension work are not rationally used. These facts were largely unrecognised by the international community prior to the World Conference on Agrarian Reform and Rural Development (WCARRD) held in 1979 which declared: "Rural development based on growth with equity will require full integration of women including equitable access to land water and other natural resources. inputs and services and equal opportunities to develop and employ their skills. There is also an urgent need to expand knowledge and statistical data on all aspects of women's roles in rural activities".

Following WCARRD there has been greater emphasis on the reduction of poverty and improved nutrition and on increased self-reliance in rural areas particularly at the household level. To this end there have been efforts to increase awareness among decision-makers and planners of the role played by women in agricultural production marketing and rural development and to better identity and respond to women's needs for training and access to inputs and services for food production and associated activities.

It is now fully recognised that agricultural extension services suffer from a number of weaknesses in their programmes for rural women. These include:

(a) a focus on efficiency objectives and few 'progressive' farmers to the relative neglect of resource-poor term households women as heads of rural households and ladles households;

(b) misperceptions and prejudices about women's actual and ideal roles with the result that they are often excluded from the target group of extension in areas in which they play major roles;

(c) attempts by separate women's units staffed by people without agricultural backgrounds to implement technical agricultural projects;

(d) a gender bias among extension workers; and

(e) inappropriate extension methodology for reaching rural women.

Historically scientific knowledge and technical skills of any given population or society spread significantly through formal and non-formal education systems. However this does not consistently happen among rural women. In agricultural countries complex socio-economic socio-cultural and religious factors need to be analysed and understood for developing effective extension and training programmes to reach women farmers.

What is the actual situation and what are the problems and solutions related to improving the effectiveness of agricultural extension programmes for the benefit of women in agricultural countries? What strategies and lines of action can be recommended to governments and external development agencies so that present and future generations of rural women in agricultural countries apply scientific knowledge and use improved agricultural techniques and skills in their farm work? These are questions that need to be answered through this proposed research.

There is some evidence to support the contention that small farmers in general, and women farmers in particular, are not being well served by the existing extension systems in developing countries . A recent FAO study of extension services in Asia and Africa concurs with this observation, concluding that: "in very few countries has an adequate organisational solution been found for advising rural women who not only work at home, but also in the fields". The failure to reach women farmers is part of an overall problem related to a lack of support and resources. In looking at women farmers as a neglected segment of the farming community, the realities of life in many less developed nations must be realistically addressed. With some extension workers being assigned to work with as many as 1000 farm families, the need for policy review is obvious. Just as obvious is the need to remember that each country has its own specific problems.

Generalisations offer much food for thought, but little basis for action. Specific situations need to be reviewed and perspective action on a country-by-country basis needs to be taken. The important contribution made by women in agriculture justifies the necessity to make the system more equitable. The above background, therefore, provides the necessary basis and justification for this research study.

WOMEN'S ACCESS TO EXTENSION SERVICES

Africa

Technical Centre for Agriculture and Rural Cooperation reported that women in Africa make up more than one-third of the work force. They account for 70% of agricultural workers, 80% of food producers, 100% of those who process basic foodstuffs and they undertake from 60% to 90% of the marketing. Further, they come up against all sorts of difficulties in their attempts to be seen as agents for development in their own right. First and foremost, they have no land ownership rights.

In some communities, they have only annual rights of use of individual fields given to them by the head of the household. Studies have shown that the majority of rural women obtained their farm land from their husbands or his family. Often this land is "given" to the women for a short period, perhaps just one growing season.

In general, they cannot make any long-term improvements to the land, such as planting perennial fruit crops, arranging irrigation facilities, etc. In 84% of water and land conservation schemes through which land has been reclaimed, only six per cent of women have been able to acquire official ownership. It was also stated that women do not usually attract the attention of credit brokers since the loans they request are often small and are costly to administer. Often credit systems use existing village cooperatives of which usually only the male heads of households are members, even though there is no formal bar to female membership.

The final obstacle to obtaining credit for women is that illiterate women cannot complete the necessary bank loan forms. Although these days African primary schools have as many female pupils as male, the rate of illiteracy among women is still above 90% in 28 African countries. All the project training tends to be oriented towards crops traditionally grown by men.

Timetables take no account of women's chores such as looking after children, cooking, cutting wood and fetching water. In many countries cultural or religious factors play an important part in preventing women from receiving training. Trainers and agricultural extension agents are usually male and thus may not speak to, or get close to women. This is especially true in Muslim countries. The international Labour organisation (ILO) estimates that 78% of women in SSA and only 64% of men are economically active in agriculture.

Food production there has long been recognised as primarily a woman's activity, but women also participate increasingly in other agricultural activities such as processing, cash cropping, animal husbandry and marketing. Women in West Africa play a pivotal role in agriculture, providing most farm labour and making the key decisions for many agricultural activities which can be documented from different studies conducted in various states of Nigeria. Ohuegbe observed that women farmers contribute more to food production and family labour than men. It was estimated that over 95% of rural women are small-scale farmers who produce most of the food and bear the burden of day-to-day family subsistence.According to diagnostic results of the agricultural extension zones, women perform almost all the cultural operations in food production. Such operations as bush clearing and burning, ridge/mount making, planting, fertilisation, weeding, harvesting, storage, processing and marketing are carried out by women. Women also have sole responsibility for cultivating compound farms (or gardens) where continuous cropping is done with household refuse.

Asia

In Asia, women provide much of the labour for the staple crop, rice. An in-depth study from Nepal gives an unusually detailed picture of the labour contribution by sex for rice and other staples. It reports that women perform 66% of the labour involved in planting, 75% in weeding, and all of the cleaning and storage of rice. In the production of wheat, they contribute 66% of the work; for maize, 94%; for oil seeds, 85%; for millet, 94%. In addition, they make 42% of the agricultural production decisions and are most influential about seed selection and fertiliser use.

In Thailand, the number of women engaged in agriculture was 6.65 million. They are the major source of labour in subsistence agriculture (100%) and are engaged in various aspects of agricultural production, animal husbandry (50%), inland fisheries (90%) and food processing (100%) and 70% are in farm management. The majority of women are engaged in agricultural work on the family farm. Seven out of ten are unpaid family workers and less than 10% are employed in private enterprises. Women generally make decisions on household expenditure and savings and some make decisions regarding big investments, such as buying land. However, men not women decide on the purchase of agricultural inputs, such as seeds, fertilisers and pesticides. The fact that women play a crucial role in agricultural production and trade has been widely ignored by development planners.

Women have been excluded from training programmes on modern methods of crop cultivation, food production, labour-saving technologies, livestock and poultry management, small-scale industries, marketing and services. Credit for technological improvements in agriculture is seldom made available to women. Membership in cooperatives through which agricultural loans are generally channelled is restricted to "heads of households" who are traditionally defined as male. It was noted that in Thailand "most technology transfer activities in agriculture, do not differentiate between men and women in their target groups. However, it is interesting to note that few women participate in the training programmes.

The government officers claim that women are not interested in new technology while according to non-governmental agencies, the number of women candidates always exceeds the capacity of their training budget". This was corroborated by women farmers at a workshop on: "Directions for strengthening the role of women farmers". There, most of the women participants expressed an interest in receiving training in new technology that would improve their agricultural productivity and lessen the burden of carrying out their tasks. It should be noted also that there are cases where government officers go out of their way to encourage women farmers to participate in agricultural activities. For example, in the dairy project in Surin Province, the majority of the group members are women.

The provincial Animal Husbandry Development board recognised the pivotal role women could play in raising dairy cows, and supported them in this activity. Income from dairy cows has become the main source of family income. With the support of the Department of Cooperatives, a Dairy Cows Cooperative was set up to strengthen the members' knowledge of marketing strategies. Provision of credit to women farmers is still limited in Thailand. In general, because poor women do not have assets which they can put up as collateral for loans, they are denied access to credit. Television has superseded all other information channels in importance regarding agricultural information for Thai farmers across the country.

Thai women and men work side-by-side in nearly all agricultural activities, with the exception of some laborious work such as land preparation. The actual hours women put into crop activities are about 64% of those contributed by men. Furthermore, it was seen that meetings with agricultural extension officers for women are uncommon, as only 20% of women attend such meetings at an average rate of two to three times per year, as compared to 60% of men attending these meetings at the rate of about five times per year.

Another interesting point is that formal sources of information are more accessible to men than to women, while women rely more on informal sources such as neighbors or their spouse. It was found that men took a more active part in farming decisions than women. With respect to household and social expenditure, women reported that they had more control than men. It was also reported by the Farming Research Institute that farming decisions were mainly dominated by men, while women had a more decisive role in domestic tasks.

Throughout South-East Asia and China, women are also heavily involved in agriculture. Within the Indonesian nuclear family, women tend to play an important role in decision-making, but outside the household they are generally subordinate to men. They are responsible for providing family food, collecting water and fuel, generating household income, providing labour in exchange for a share of the crop and generating an independent income from wage labour, handicrafts, agro-processing and marketing. Philippine women are noted for their many entrepreneurial activities, often in the informal sector and independent of their husbands.

South Asia showed that women from large landholding families who supervise farm labour share their agricultural information with the women laborers, with beneficial results for both the landowners and the laborers. In these areas, landed women would make good contact farmers. Data from commercial banks in Andhra Pradesh, India, showed that only 6% to 12% of the loans were given to women and, moreover, none of these were agricultural loans. A survey of households in Bangladesh found that women made up only 2.8% of the recipients of formal credit.

Latin America

Estimates from Peru and Chile, showed rural women's contribution to the work in agriculture as 21 % and 10.4% respectively. Some suggested that the primary impact of the Western influence on gender roles in a society is to transfer much or all of the non-farm subsistence work to women, followed later by a transfer of subsistence farming. Case studies indicated that in many cultures, men clear the land and do the initial ploughing, whereas women plant, weed and help with the harvest. An interesting version of this division of labour is seen in Jamaica where, on steep hillsides, men prepare the soil with a plough for both subsistence and commercial crops. The tasks that women do also vary according to farm size, or when implements are involved. On larger farms where many modern inputs are used, female participation is limited, but on the smaller farms there is a higher level of sharing of tasks and decision-making between the sexes. In general, as shown by field surveys in Trinidad and Tobago, women farmers perform the less strenuous tasks such as planting, cutting, weeding, fertilising, moulding of soil around young plants and harvesting. Some of these tasks are gender-neutral or interchangeable, especially harvesting and fertilising.

Pest control is less likely to be undertaken by women because it is thought that the use of chemical sprays is dangerous to women, especially if they are pregnant or lactating. Of the 96 farmers in the Trinidad survey who used insecticides, weedkillers and fungicides, men performed the task on 77 farms, whereas it was jointly carried out on 10 farms. Only five females, all of whom owned their own farms, sprayed their own crops. On other female-operated farms, pest control was delegated to hired or family labour.

Some argues that Jamaican women play an important role in food production as small producers and as agricultural labourers. In an agro-socio-economic survey conducted in six parishes in Jamaica, 45 % of agricultural producers were women. The Agricultural Census of Jamaica showed that, of the total of 182,169 farms island-wide having single holders, 19% were operated by women.

Near East

Rural women in Syria do a great deal of agricultural work. They are more involved in lighter work that does not require great physical effort, but needs care and patience: planting, transplanting, weeding, thinning, threshing, harvesting and pulling out the roots. Such activities usually continue throughout the year and are often of a more tedious nature. Most of the work is done manually or with the use of hand tools. Poultry raising, care of livestock, milking and milk processing at the family level are also the responsibility of rural women. It was also noted that illiteracy among females engaged in agricultural work was more than 90% and they were mainly considered unskilled cheap labour. Available information has shown that the redistribution of land under the agrarian reform programme did not significantly increase the number or proportion of women landowners in the country. Membership in agricultural cooperatives is open to male and female farmers; nevertheless, social cultural barriers still limit rural women's participation in cooperatives. It was reported that female membership is increasing, but at a very slow rate, i.e. 0.5%.

In most of western Turkey, women are much involved in agriculture, particularly for the production of fruits and vegetables and the rearing of livestock. There are fewer taboos against men and women interacting than in parts of eastern Turkey, where women's roles reflect traditional Islamic norms. In eastern Turkey, women are still active in farming, but more in the seclusion of their homesteads.

In Yemen, male emigration is particularly heavy, where an eighth of the population works in paid employment elsewhere. The women let's behind have had to assume the agricultural duties of the absent husbands and, as direct recipients of remittances, have greater decision- making power than would be expected in an Islamic country. About 80% of women live in rural areas and work all the year round. In contrast, men mostly work seasonally on heavy work, technical (pesticide spraying) work, marketing and delivery, and cash crops.

REFERENCES

Bne Saad, M. H. A. Al-h. 1990. *An analysis of the needs and problems of Iraqi farm women: Implications for agricultural extension services.* Unpublished doctoral thesis, University College, Dublin

Hayward, J. 1990. Agricultural extension: The World Bank's experience and approaches. In FAO *Report of the Global Consultation on Agricultural Extension* (p. 115-134). Rome: FAO.

Rivera, W. M., & Gustafson, D. J. (Eds.). 1991. *Agricultural extension: Worldwide institutional evolution and forces for change.* Amsterdam and New York: Elsevier.

11

Innovations in Agricultural Extension

Agricultural extension organisations, like most other disciplines and institutions, are not immune to various developments taking place around them. Some major developments and their influence on various aspects of extension are mentioned here.

IMPACT OF GLOBALISATION AND LIBERALISATION

Globalisation aims at creating more interaction and linkages among countries in trade, information flow and finance, encouraging open competition through removal of trade and other national safeguard barriers, supposedly under fair and relatively equal conditions for producers of goods and services no matter which country they live in, thus making it possible to mutually benefit from international contacts and opportunities.

The controversy, however, lies in the uncertainty of fair and equal conditions for all producers, irrespective of the advantages, disadvantages, potentials and risks of globalisation. Under present conditions, the developing countries cannot compete with the developed nations in the international market in terms of production, quality and exports because of a number of factors such as lack of resources, lack of technology, weak rural institutions, poor infrastructure and communication facilities, etc. The main message transmitted in the by-now familiar demonstrations at economic summits is that globalisation will make rich countries richer and poor countries poorer.

The WTO conference held in September 2003 in Cancun, Mexico saw serious disagreement between rich and poor countries on agricultural issues, and the event ended in miserable failure. Apart from politics, the phenomenon of globalisation will certainly bring both risks and opportunities to the farming communities of less-developed countries.

The communities, therefore, must be educated and prepared to modify their agricultural operations within the context of globalisation, a responsibility which, by and large, the national agricultural extension systems will have to bear, and they must prepare themselves in time to meet the challenge. Market liberalisation is an integral part of globalisation. This calls for removal of trade barriers, tariffs and other regulatory measures that are usually put into force by countries to protect their own industries and products, and to discourage the inflow of articles produced elsewhere in the world.

Liberalisation asks for the opening of markets, called deregulation, so that goods may move freely between countries. It also advocates the removal of artificial price controls and of public support to the farming sector in the form of farm subsidies so that the market can realistically determine the price of various commodities and products on the basis of their demand, supply and quality, and the consumers can then freely make a choice concerning what is best for them. However, at present, heavy government subsidies in rich countries give a huge advantage to their farmers over the small and poor competitor farmers of less-developed countries, threatening their very survival. For example, small farmers in Mali who have been growing cotton for centuries as their main source of the income may be wiped out if cotton, produced at a low cost in developed countries as the result of government subsidies, continues to be dumped and sold cheaply in the Malian market. Both commercial and subsistence farmers in developing countries are bound to be affected, directly or indirectly, sooner or later, by market liberalisation, and national agricultural extension services should be knowledgeable enough to educate the farmers in how to properly enter a liberalised market.

Millions of subsistence farmers produce barely enough for their own consumption; in a very favourable cropping season, they may produce a bit of surplus to be sold at market. However, today there is an all-out force aimed at commercialisation of farming even on a small scale. There are both ethical and technical questions, about whether it makes sense to let the subsistence farmers continue as they have been doing for generations, or to allow their operations to be transformed into agri-business, rural enterprise, rural industries or other commercial ventures, or whether very poor farmers of remote, marginal areas should be encouraged to migrate elsewhere because investing in development for them is not cost-effective.

Appropriate agricultural policies and land consolidation laws will be required if privately owned and cultivated small units of land are to be combined or merged to create larger, commercially viable plots. The agricultural extension services will have to come up with the strategies that could help subsistence farmers in organising themselves for commercialising their operations profitably, without losing the pride of land ownership. A global debate is raging on the issue of producing and consuming genetically modified organisms (GMOs). There are various positions which represent different points of view.

On the one hand, all possible means should be used to feed increasing populations, but on the other hand, biodiversity must be protected, and people must be alerted about the health risks to humans and livestock potentially resulting from genetically modified diets. In this debate, the most ignorant are the farmers in developing countries who are growing food for millions of people. No agricultural extension services so far have felt the need to educate farmers about this controversial but important subject. Since biotechnology is not yet widespread in developing countries, there is no urgency to start any special extension programmes. But this is precisely the time when necessary steps should be taken for preparing training courses and teaching materials both for extension workers and farmers.

IMPACT OF PRIVATISATION

Many services that were managed in the past by governments are now being managed and delivered by the private sector, especially in developed countries. Many developing countries are following the suit. The underlying reason is dwindling budgets of public institutions, which makes them relatively inefficient and less productive, causing not only financial loss to the government but also creating discontent among people. The private sector, on the other hand, has generally more resources, innovative ideas, and a motive for profit and is thus keen to offer efficient and better services to its clientele.

In recent years, several multilateral and bilateral donors have encouraged the privatisation of national extension services. Many arguments are put forth in favour of privatisation, e.g. public extension systems have failed or are at best unsatisfactory; since farmers profit from extension advices and the government budgets for public services are decreasing, the cost of delivering advice should be recovered by charging a fee from the farmers; the private sector is actively involved in selling farm inputs and machinery, thus why should it not handle the task of advising farmers on agricultural matters; healthy competition among service providers will lead to better quality and lower costs for services; it is prudent to consider financing and delivery of extension services as separate functions, to be performed by two different sectors, i.e. public and private.

The trend to privatise extension services will certainly shake the traditionally friendly and informal relationship between government extension staff and the subsistence farmers in developing countries, because until now the farmers have never been asked to pay for extension advice. Apart from the question of whether total, partial or no privatisation of extension is needed in developing countries, the national extension services and their clientele should be well versed in the pros and cons of this major issue. Several developed countries have fully or partially privatised their agricultural extension services in a variety of ways. Terms like outsourcing, cost-recovery, and contracting out are related to the drive for privatisation. Costa Rica has a unique system under which the government provides farmers with extension vouchers which can be used for getting extension advice from private specialists.

In England, the public extension service has evolved over time into a private consulting practice. The positive result is enhanced efficiency of staff, and the negative effect is the deprivation of small farmers of extension services as the result of their inability or unwillingness to pay. It is also said that the government has taken over some previously privatised advisory functions because of dissatisfaction with the private sector.

In Holland, about 60% of the extension budget comes from farmers, while the remaining 40% is provided by the government. The benefits include increased efficiency, improved quality, client-orientation, job satisfaction for staff, and expanded marketing opportunities for farmers. The problems faced include loss of government authority, the government's inability to keep financial promises, and weaker communication with the stakeholders because of creation of competition among them.

In Albania, the private sector entrepreneurial initiatives to create a long-term relationship with farmers have proved to be successful. The extension services in Nicaragua are both decentralised and semi-private. Bulgaria privatised a number of state farms to be used as demonstration farms, with an objective of establishing private extension service.

Since the experiment was not successful, the government decided to establish a National Agricultural Advisory Service (NAAS) with external financial assistance. The NAAS is now being reviewed, with FAO assistance, for possible reform in order to meet new learning needs of farmers, as the country is scheduled to join the European Union in 2007 or soon thereafter. Estonia has both a public extension advisory service for poor farmers and a private service for better-off farmers.

Uganda is experimenting with the privatisation of extension through the creation of a pool of private extension specialists out of its existing public extension service; registered farmers' associations could call upon this pool through bidding for providing services related to selected enterprises, and pay for the services from the funds given to them by the donors through decentralised government units. It is indeed a bold experiment, but the sustainability of this arrangement remains to be seen after the donors' funding runs out.

In Israel, the efforts to even semi-privatise national extension services have not met with success. The government is still responsible for providing extension advice, but encourages privatisation through the standing practice of growers to contribute portion of their income to research and development including extension, public and private partnership in financing and operating units within the extension service, payment for services by commodity production and marketing boards beyond a basic extension package, the provision of more intensive extension activities at the request of needy growers, special agreements with commodity farmers' organisations, extension staff working on their day off in exchange for direct payment from farmers, provision of equipment like mobile phones to extension advisers by growers associations, and direct payment by farmers for participation in training activities.

The advocates of privatisation of extension services believe that farmers should pay for the extension advice. However, there is genuine fear that the zeal for privatisation would deprive small farmers from benefiting from the services. The small farmers either do not believe that the extension advice is worth paying for, or they simply cannot afford to pay.

The common wisdom would dictate that in developing countries, commercial farmers and large cooperatives should pay for extension advice, while the government should continue providing free-of-charge extension services to small producers. Measures are also needed to protect the farmers from exploitation by the private sector.

IMPACT OF PLURALISM

The spread of privatisation has shifted many sectors of public services from the government to private companies. Government monopoly is not the most favoured option in the developed countries, especially in matters of development when competent non-public institutions are present in the country. The modality of using more than one organisation, whether public or non-public, for delivering extension services to farming communities, is gaining popularity. The obvious rationale is the pooling of all available resources in order to alleviate pressure from low budgets and staff in the ministries of agriculture, as well as to let the farmers benefit from a variety of sources. Many developing countries are already practising pluralism in extension.

In Zimbabwe, many NGOs, private companies and semi-autonomous bodies are to be found engaged in delivering extension advice to farmers. Bangladesh's national policy on agricultural extension recognises, for the first time, the need for partnerships with NGOs and the private sector under a decentralised extension system.

In Honduras, where extension services are being privatised and small farmers are unable to pay, 70 NGOs reach about 50,000 farmers living mostly in remote areas. In Indonesia, some projects have not only encouraged NGOs and the private sector, but also agricultural research institutes, agricultural universities and farmers' associations, to participate in the delivery of extension services.

The main challenge in installing a proper pluralistic agricultural extension mechanism is the effective coordination among various agencies, the absence of which could lead to conflicting technical recommendations and creating confusion among the farmers. The governments should take the responsibility for coordination, technical supervision and quality control.

DECENTRALISATION AND DEVOLUTION OF EXTENSION SYSTEMS

Decentralisation and devolution have been adopted in many countries and the trend continues. The main rationale for the move is that decentralised administration shifts decision-making authority to lower administrative and political level units, which—being physically close to people—are supposedly in a better position to plan and implement development programmes with the active involvement of participants.

The concept is sound, but its implementation in developing countries has so far not been smooth for various reasons. For example, some central governments are reluctant to relinquish their control over decision-making, especially in financial matters; there is weak capacity of the decentralised units in running government affairs; and the unwarranted interference of local politicians in technical matters in which they are not competent hinders implementation.

Decentralisation has, in many instances, placed agricultural extension responsibility in the hands of the ministry of local government. Although decentralisation is good in principle, the initial stage shows quite a bit of setback for extension. In case of Philippines, for example, the interference of municipality-level politicians has compromised the effectiveness of extension programmes to the extent that persons without diplomas in agriculture have been recruited against agricultural extension vacancies. Also, the career development opportunities of extension staff have been adversely affected.

The latest thinking is that since the municipalities are not viable economic units for delivering extension services, the extension responsibilities should be moved back to the provincial level. In Indonesia, extension services have been marginalised because most district governments have certain priorities, irrespective of national policy, which could generate quick revenues, and agricultural extension which, unlike estate crops or livestock, generally gives a long-term return, is not considered as a priority.

United Republic of Tanzania's decentralised extension service at field level suffers from lack of operational budget. In Kosovo, where the government comprises UNMIK (United Nations Interim Administration in Kosovo) expatriates, the decentralised government units do not want to do much with the central government. In Uganda, some district authorities have preferred to spend the extension budget on constructing feeder roads, leaving extension staff without salary for several months.

One reason for this "institutional chaos", which applies to many countries passing through decentralisation and devolution, is that the central government did not adequately prepare itself or the local governments in advance before embarking on decentralisation. Decentralisation is truly a double-edged sword, and if not handled properly, could cause more harm than good. As more and more countries come under decentralisation, their extension services are going to face the problems of marginalisation and lack of professional identity unless they are prepared well in advance for this critical administrative, technical and physical transition.

CLIENT PARTICIPATION IN DECISION-MAKING

Civil society advocates ever more strongly the reasons why democracy, participatory decision-making, transparency in government affairs and good governance are necessary for eradication of poverty and hunger, uprooting corruption, relatively equal distribution of benefits among various sections of society, welfare of disadvantaged and vulnerable groups, optimum utilisation of human and physical resources, sustainable livelihoods, and overall human development. Even the economic aid given by bilateral donors to poor countries, in many instances, has depended on the government's records on human rights and good governance.

A large number of strategies have evolved during recent years, which supposedly ensure participatory decision-making and involvement of all stakeholders in joint planning and implementation. All grassroots-level service institutions have been influenced by these conceptual thrusts. The powerful trend towards empowering farmers through active participation in decision-making has led to various emphases in extension, such as working through farmers' groups, preparation and delivery of client-oriented messages, observance of gender sensitivity, and research-extension-farmers linkages. Certain participatory tools like PRA (participatory rural appraisal) and KAP (knowledge, attitude and practice) survey have also been developed.

Indonesia has successfully established new institutions called Agricultural Technology Assessment Institutes at provincial level, bringing together farmers, researchers and extension specialists.

In Pakistan, groups of highly motivated small farmers established under FAO Special Programme on Food Security have been taking decisions in matters related to group cash savings, quality seed, fertiliser, water-management, cultural practices, farm machinery, income diversification and marketing of produce. The pilot experiment has proved to be so successful in terms of yield increases that the government has embarked upon replicating the approach in thousands of villages. In the northern region of Pakistan, FAO has introduced a demand-driven, bottom-up, gender-sensitive extension service that involves men and women villagers' groups in preparing grassroots demand-for-services plans.

In Argentina, one main factor for success of a federal programme for small- and medium-size farms is the involvement of all stakeholders in major decision-making. Similarly, Jordan's extension service has established Close Contact Groups of farmers while United Republic of Tanzania has formed Participatory Farmer Groups of men and women farmers. In the Philippines, where extension services have been devolved, an

FAO project has assisted citizens in outlining a methodology for participatory, grassroots extension programme development and in preparing an extension delivery partnership mechanism that involves stakeholders such as extension staff, farmers, NGOs, private sector, research institutes and academic institutions.

IMPACT OF NATURAL AND MAN-MADE DISASTERS

Human miseries are unfortunately on the rise in an age of civilisation and scientific progress, some man-made and some natural disasters. There are conflicts, wars, droughts, famines, storms, earthquakes and epidemics, all of which take an enormous human and physical toll and disrupt the normal course of life. There has been a steep increase in the number of countries afflicted by human-induced disaster, i.e. from an average of 5 in the 1980s to 22 in 2000, mainly because of armed conflicts. Between 30 and 40 countries were engaged in conflict at the end of the twentieth century, adversely affecting thousands of millions of people.

Farmers of some countries like Afghanistan cannot farm because of the land mines in their fields. Basic farming ingredients like seed, water and soil are not readily available. Extension services have dissipated and the farmers have no source of technical advice. Infrastructure has been damaged, making the transportation and delivery of farm inputs impossible. Although food aid agencies and NGOs have been active in disaster-hit areas, food handouts work only up to a limit. In recent years, weather-related and other natural disasters have risen from 10 to 18 incidents per year. In 1998 alone, some 32 000 persons were killed and 300 million were displaced from their homes and livelihoods.

The tsunami of December 2004 killed hundreds of thousands of persons in a number of countries and wiped out or seriously damaged farming infrastructure in the coastal zones. The very strong earthquake that mainly struck the Pakistan-administered Azad Jammu and Kashmir region in October 2005 killed over 50 000 persons, mostly small farmers of the mountainous villages, and destroyed physical infrastructure and livelihoods.

In some countries like Bangladesh and the Caribbean where huge storms hit every year, extension services are being trained to provide assistance in post-disaster situations, but it is true that agricultural extension services have yet to play a significant role in the rehabilitation of agricultural operations after the disasters. It is realised that extension services cannot face such colossal challenge by themselves, but they should work in collaboration with relevant institutions and help the rural population in growing food for survival using whatever resources are available. A comprehensive response from national agricultural extension systems is still awaited.

IMPACT OF INFORMATION TECHNOLOGY REVOLUTION

We are in the midst of an information technology revolution which has virtually shrunk the world and has affected almost every walk of life. Information technology is indeed a tremendous power that could be harnessed by organisations for the benefit of humanity, and extension services should not hesitate to utilise this power. They can exploit this potential to strengthen their own capacities and to educate the rural populations who enjoy access to media.

The extension organisations in developing countries have two major problems when it comes to having face-to-face contacts with the farmers and researchers: first, physical distances and the second, lack of transportation facilities. The new information technology could bypass these physical barriers to a great extent through the development and application of appropriate, interactive information mechanisms. Advanced information technology is already making headway in the area of rural and agricultural development.

A number of developing countries such as Lao People's Democratic Republic, Viet Nam and Mali, are experimenting with telecentres, which have already exhibited their benefits in several Western European countries. Virtual linkages are being established for bringing research and extension together, and one example is the VERCON (virtual extension, research and communication network) tool, which FAO has introduced in Egypt and Bhutan. Under an FAO project in the Philippines, the Internet and interactive e-mail facilities have been established at municipality level for supporting decentralised extension staff. Expert systems are also being developed to compensate, to some extent, for the too-rare visits of subject-matter specialists to farmers' fields.

The use of cellular phones is by now a routine practice and the equipment is used for rural development projects in Bangladesh. Over 30% of extension staff in Estonia use the Internet. One can find programmes like "virtual gardens" and "virtual farms" on the Internet. The main issue is how the powers of advanced information technology can be harnessed for the benefit of both extension agents and farmers without compromising the importance of unique local factors such as indigenous communication patterns, and also without considering information technology as replacement for the extension agents, which remains a much-needed and -appreciated human element.

IMPACT OF RURAL POVERTY, FOOD INSECURITY AND HIV/AIDS EPIDEMIC

There have been internationally renewed calls, the latest coming through the Millennium Development Goals, for eradication of rural poverty and hunger, and for support to vulnerable groups such as the rural landless, women and youth. The number of poor has gone up in recent years. Moreover, even though there is enough food produced for everyone in the world, there are still over 800 million people—almost one in seven persons—who do not have enough to eat.

Hunger, poverty and vulnerability are inter-linked. While the world celebrated the new millennium, 13 million people reached the brink of starvation in the Horn of Africa, and European farmers and their livestock industry were hit by mad-cow and foot-and-mouth diseases,. followed by the avian flu epidemic in some Asian countries and which has now spread to Europe. In 2005, when the G-8 leaders were pledging billions of dollars for development of Africa, a vast famine struck Niger, killing thousands of persons including many children. According to the estimates from the UNAIDS/WHO AIDS Epidemic Update, 37.2 million adults and 2.2 million children were living with human immunodeficiency virus (HIV) at the end of 2004; 3.1 million died that year alone from AIDS.

The overwhelming majority of people with HIV (some 95% of the world total) live in the developing world. Sub-Saharan Africa is by far the worst affected region in the world by the AIDS epidemic. The region is home to just over 10% of the world's population, but has over 60% of all people living with HIV. An estimated 3.1 million adults and children became infected with HIV during the year 2004. Around 1.2 million people in Asia acquired HIV in 2004, bringing the number of people with HIV to an estimated 8.2 million.

Alarming figures have been reported for other regions of the world. There are indisputable, negative effects on the workforce resulting in the loss of trained, skilled and experienced workers in all disciplines. Farm labour, plentiful in the past, is rapidly diminishing. As studies such as those carried out by FAO in Malawi, Uganda and Zambia show, the epidemic has also affected agricultural extension organisations in the sense that not only there have been deaths and long absenteeism among the staff, but the old, traditional extension approaches have also been rendered unsuitable. This is due to the fact that the labour-intensive cropping patterns have been changed as a result of the weakening physical condition of farmers and lack of farm labour and the emergence of "new farmers" comprising the elderly, widows and young children; the unsuitability of current farm tools; and the unsuitability of existing rural credit approval criteria. Extension organisations in the countries affected by HIV/AIDS, especially in sub-Saharan Africa, must create new extension strategies supported by modified farming systems and appropriate farm tools.

New anti-hunger initiatives are under way, especially since the 1996 World Food Summit in Rome, involving all the institutions that deal with rural populations. Similarly, the New Partnership for Africa's Development (NEPAD) has been launched. It would be difficult for agricultural extension services in frequent contact with rural people to remain isolated from these international efforts against rural poverty, food insecurity and the HIV/AIDS epidemic. Extension services must develop capacity and suitable extension educational strategies and tools to effectively participate in the efforts against these human miseries.

SUSTAINABLE DEVELOPMENT AND EXTENSION SYSTEMS

The global preference for multi-disciplinary, integrated and holistic approaches to development is now an undisputed fact. The rationale is that simultaneous, multi-sector development is more meaningful than individual sector development. More and more organisations are revising their structure in favour of this choice. The "integrated rural development approach" which was applied during 1960s in many countries is returning in a revised form, including elements of bottom-up planning, participation by stakeholders, emphasis on eradication of rural poverty, and gender-sensitive approaches.

Many multilateral donors have adopted a programme approach, abandoning the old project approach. FAO has run an integrated programme in Namibia, Uganda, Zambia and Zimbabwe, with as many as six technical components including research and extension. The sustainable livelihoods concept also promotes holistic development.

The Neuchatel Initiative Group comprising major donors interested in reforming national agricultural extension systems has been exploring the role of extension in a wider rural and agricultural development context. FAO and the World Bank have been working jointly on AKIS/RD (Agricultural Knowledge and Information Systems for Rural Development), which is aimed at increasing cooperation and joint operations between research, extension, education and farmers.

FAO has conducted studies on AKIS/RD in ten countries, in various regions of the world, which were later analysed and synthesized with the objective of learning lessons and making recommendations for bringing AKIS/RD actors closer to one another. There still exists a temptation among various agricultural line departments to maintain individual extension services, but as problems with this approach have been observed and well-documented, extension services are being unified in a multi-disciplinary manner in the interest of optimum utilisation of resources and an efficient bureaucracy. Indeed, the farmer's time cannot and should not be wasted through individual visits of many extension agents, each representing a different agricultural discipline. The creation or strengthening of multi-disciplinary subject-mater specialists' teams during decentralisation of extension services in a number of countries is a popular move.

FAO has provided technical assistance to Uganda in the integration of agro-forestry and HIV/AIDS education in the agricultural extension programmes, thus making the extension approach multi-disciplinary. In Iran, FAO, while assisting the government in restructuring the national extension system, has recommended the placement of multi-disciplinary teams of subject-matter specialists at district level, the composition of each district team to be determined by the technical needs and priorities of the particular district. Thus the districts in coastal zones will have a fisheries specialist, and those with dominant livestock activities will have a livestock specialist.

In Indonesia, both agriculture and forestry extension have been placed under a single extension service assisted by a World Bank-financed project. The same has been done in Lao People's Democratic Republic. In the Philippines, extension service covers both agriculture and fisheries under the Agriculture and Fisheries Modernisation Act. Under a project funded by the Asian Development Bank, Viet Nam's extension services are being reformed, and one of the reform measures will include unification of the present extension services.

The promotion of sustainable development asks for the integration of certain non-agricultural but very relevant elements into extension messages. FAO has been assisting Egypt in integrating population and environment education messages into ongoing agricultural extension programmes. Similarly, Pakistan has been provided assistance in preparing extension services to play a key facilitation and participatory role in educating rural people in managing natural resources properly. FAO's integrated pest management programme (IPM) has been promoting biological control of insect pests for years, thus discouraging excessive application of chemical pesticides.

REFERENCES

Russell, E. J. 1966.. *A history of agricultural science in Great Britain, 1620-1954.* London: George Alien and Unwin.

Smith, H. 1972.. *The Society for the Diffusion of Useful Knowledge, 1826-1846.* London: Vine Press.

True, A. C. 1928.. *A history of agricultural extension work in the United States, 1785-1923.* USDA.

Zijp, W. 1994.. *Improving the transfer and use of agricultural information: A guide to information technology.* World Bank Discussion Paper 247. Washington, DC: The World Bank.

12

Modernisation of Extension Systems

In recent years, the national public extension systems have been increasingly criticised for being not that effective for all the investment that has gone into maintaining organisation and staff. They have been called a burden on public funds and, as such, have been targeted either for total abolishment or for drastic reforms. Certain extension services have been said to be so large that they are unable to move efficiently because of their own weight.

Some of this criticism is genuine and some is not, when the financial, logistical and career development constraints under which extension workers are forced to operate are examined objectively, and when these services are compared to other disciplines such as agricultural research with far better service conditions.

The fact remains, however, that modernisation and reforms are needed in the existing national extension systems as a result of the many global forces that are changing socio-economic and political conditions in the world, creating new challenges and learning needs for farmers in developing countries.

ASSESSMENT OF EXISTING EXTENSION ORGANISATIONS

All organisations are built and filled with human resources to serve specific purpose within the context of a policy and mission. Public institutions are funded with the public funds and as such are supposed to serve the public. In case of agricultural extension, the organisation is meant for serving the extension, education and training needs of men and women farmers. If it is established that the organisation is not delivering effectively and efficiently what it was built to deliver, then it should be reviewed. A quick way will be to assess it on the basis of extension and training needs of the farming communities.

Key Actions

– Assessment of extension and training needs of farmers:

  - Preparation of a survey tool, such as detailed structured questionnaire
  - Preparing data-collectors through brief training
  - Selection of survey sample areas
  - Selection of men and women farmer categories sample comprising subsistence, commercial, special interest, landless, etc.
  - Survey for data collection on farmers' responses to the questions through individual and/or group interviews
  - Analysis of data, identification of extension and training needs of different categories of farmers, and their prioritisation in light of government policies and goals of rural and agricultural development
- Keeping in view the prioritised extension and training needs of farmers, an assessment of the following with the objective of identifying specific areas in the existing arrangements which need to be revised, strengthened or restructured for meeting the farmers' needs satisfactorily:
  - Existing organisational structure
  - Operational linkages with relevant institutions
  - Present mission and technical mandate
  - Human resources
  - Physical facilities
  - Operational budget
  - The presence or absence of competent or relatively competent non-public institutions including NGOs, farmers' associations, rural community organisations and private companies with adequate staff, interested in delivering extension services under contract with the government
- Keeping in view the gaps identified as a result of the assessment of the above components, as well as the results of farmers' extension needs survey, make a decision whether it will be more cost- and result-effective:
  - to continue the public extension organisation as it is;
  - or to modify it as necessitated by the field survey results;
  - or to create a mixture of the existing organisation and non-public institutions
- Based on the decision on the preceding point, preparation of an action plan, financing and schedule for overall organisational modification, or strengthening or

restructuring of the specific organisational components including revision of terms of reference, with participation of the key staff. If there is evident gap between the technical specialists present in the existing organisation and the technical specialisations needed on the basis of the farmers' needs identified, assuming that adequate funding is available, then "outsourcing" option, i.e. contracting out relevant extension work to the non-public sector, should be considered rather than hiring new technical staff. This strategy will save time, money and effort.

– Implementation of the plan.

DECENTRALISATION OF EXTENSION SERVICES

While decentralisation is a step in the right direction, it has proved, at least so far, to be disastrous for agricultural extension in several countries. The issue of the temptation of each autonomous district or municipality to go in its own direction without any regard for national policy and priorities is serious, but what has damaged extension most is the unwarranted influence and interference of local politicians and the mentality of local decision-makers and legislators, who cannot appreciate the importance of extension in rural and agricultural development. This has resulted in the diversion of extension budget to other activities, assigning of extension staff to non-extension tasks, use of its few beaten-down equipment and vehicles for personal purposes, and hiring of non-agriculture graduates for agricultural extension positions.

As the local decision-makers mostly depend on central government for their financial needs, their preference is ruled by a strong temptation to earn revenue in shortest possible time in order to reduce this dependency. Agricultural extension, being educational in nature, is understandably a slow process as it aims at changing human behaviour and generally does not show quick and tangible benefits.

The local decision-makers, who are elected for two to three years term, prefer to select activities like livestock and estate crops, which give sizable, tangible and relatively quick returns, or to build roads and shopping centres, which could win voters' appreciation and guarantee votes for the next election. This tendency of local decision-makers, most of whom have limited education, needs to be changed through their orientation to national policy and priorities, proper education, sensitisation and orientation to the importance and eventual benefits of extension. Unless this important group is "converted", agricultural extension will remain marginalised, or might altogether disappear, under decentralised administration.

The countries like Nepal and Pakistan, which have recently embarked upon decentralisation/devolution, can make the process smooth by taking capacity-building measures for decentralised units, and can also draw other lessons from the experiences of countries like Chile, Colombia, Uganda, Philippines, Indonesia, United Republic of Tanzania and Ghana where decentralisation came into effect many years ago. This will help in avoiding the pitfalls of decentralisation experienced by the "senior" countries.

Key Actions

- Prepare extension staff in advance of the decentralisation through the following measures:
  - Provide the staff with revised terms of reference
  - Explain to the staff the new organisational and administrative arrangements, including topics such as relationship with the local government officials and the Ministry of Agriculture, relationship with other decentralised units, budget allocation, any major effects on staff employment conditions during transition to decentralisation, physical relocation in new offices at district or lower levels, institutional linkages at decentralised level, etc.
  - Provide training in the following technical subjects: philosophy of decentralisation; participatory extension approach; demand-driven extension; grassroots, bottom-up programme planning; organisation of men and women farmers' groups; integration of village extension plans into district development plans; participatory socio-economic impact assessment of extension activities.
- Either include a module on extension in the existing training programme for the elected officials of the local government, or organise special orientation sessions on extension. The orientation should focus on the importance of extension in bringing positive behavioural change among farmers for sustainable rural and agricultural development, even though it takes longer time than short-term investments into cash crops which give quick returns. Other topics could include importance of recruiting well qualified candidates for extension positions, terms of reference of the extension staff which show that they are not responsible for non-extension tasks,. transport facilities for extension staff to perform satisfactorily, the importance of the emphasis of extension workers on small farmers as opposed to favouring big farmers, importance of gender consideration in provision of extension services, respect for decisions made by technical subject-matter specialists, avoiding shifting of extension budget to non-extension activities, non-politicisation of technical matters, and importance of career development opportunities for the extension staff.
- Consider decentralising agricultural research at the same time when agricultural extension is being decentralised, because it has been observed that the institutional linkage problems were more prominent in those countries where only extension was decentralised and research was left centralised.

– If there has been no formal decentralisation, but the bulk of technical subject-matter specialists and the extension staff are based at national and provincial levels, then exercise "de-concentration" and move bulk of the staff to district and lower levels so that the staff, being physically close to the farmers, could plan and implement the extension activities in collaboration with the farmers. The re-location of the subject-matter specialists should be done in line with the technical needs of respective districts. For example, if a district lies next to sea or river then a fishery expert will be needed, and if the livestock activity dominates in a district then a livestock specialist should be posted.

BROADENING OF TECHNICAL MANDATE OF EXTENSION

Extension is basically non-formal education that targets rural adults outside the formal school system with the aim of helping them improve the quality of their lives by gaining useful knowledge and skills. Agricultural extension focuses on the non-formal education of rural adults, in particular farmers, in improved agricultural know-how and skills for increasing farm production, which is meant to result in enhanced income for farmers, leading to improvement in their lives. Sustainable rural and agricultural development, however, goes far beyond improved agricultural technology to enhance production, because it involves not only crops, livestock and fishery but also the protection of environment, management of natural resources, maintenance of appropriate population growth rate, and above all, development of rural human resources.

No other ministry in most developing countries enjoys the presence of such a large number of workers in the field, i.e. extension workers, as does the Ministry of Agriculture. However, the present technical mandate of this precious field staff is extremely narrow, limited to the transfer of improved agricultural technology to farmers.

As improved technologies are not generated by agricultural research institutes so frequently that the extension workers can pass on something new to farmers on a daily or weekly basis, this narrow mandate keeps this valuable stock of field workers underemployed. It is also true that no government workers come in contact with the villagers and farmers as often as do the extension workers.

Thus, the presence of this field work force provides an excellent opportunity for non-formally educating the rural residents in the knowledge and skills which could promote the concept and practices of sustainable rural and agricultural development. This, however, will require broadening of the technical mandate of extension staff and training of the staff in the concepts and technical components of sustainable rural and agricultural development. It may also require certain changes in the way various technical departments work for developing rural areas.

Key Actions

- Immediate preparation of the existing extension staff for non-formal education of farmers in the concept and practices of sustainable rural and agricultural development through provision of in-service training. The following actions will be needed:
  - Preparation of short-term training modules and materials on the following topics: concept of sustainable rural and agricultural development; interrelationship among environment, population and agricultural production; wise management of natural resources; the impact of any major epidemic, such as HIV/AIDS, on farmers and farming practices and how to mitigate it; how to farm under market liberalisation conditions; problem-solving and decision-making; managing farm as business; farmers' group dynamics for lobbying; rural leadership for development; participatory decision-making; gender sensitivity in the planning and delivery of extension services; rural youth development for healthy future; self-help initiative and use of revolving fund for development activities; basics of information technology; food security and how to ensure it throughout the year; off-farm income generation activities during slack period amidst cropping seasons; socio-economic impact of extension activities; grassroots, bottom-up planning of extension programmes; post-disaster rehabilitation of agriculture; rural poverty; food security.
  - Preparation of a training plan for provision of in-service training to extension workers, based on the prepared training modules and materials, on as many topics as practically possible, and implementation of the plan.
- Broadening of teaching programme at agricultural colleges and universities in agricultural extension through revision of curricula, books, materials, and teaching methods.
- Short-term training of men and women farmers by the extension services in the subjects of problem-solving, decision-making, management, accounting, group dynamics, leadership, participation, gender sensitivity, rural youth development, good governance, citizenship, initiative and self help, nutrition, programme planning, monitoring and evaluation, information technology, importance of education for children especially for girls, networking with other village organisations and farmer associations, etc.
- Establishment of operational linkages between the Ministry of Agriculture and those ministries and line departments, which are responsible for the technical components, which were never before handled by the extension workers, such as population, environment, HIV/AIDS, with the objective of coordinating the field work for the benefit of farmers.

– Consider changing the name of agricultural extension to agricultural and rural extension, or extension for rural and agricultural development

FORMULATE NEW POLICY ON EXTENSION

So far, only a few countries, like Bangladesh, Nepal and Philippines, have formulated a national extension policy. The existence of policy is of great importance because it ensures political commitment, which ascertains financial allocation. The dismally low salaries and operational budgets, and a lack of attractive career development path have been the fate of extension profession for decades affecting the performance of otherwise very experienced and committed professionals. The situation for extension is particularly bleak where decentralisation or devolution has occurred. Appropriate policy is indeed needed for protection of extension from immediate negative effects of decentralisation to allow it to function normally. It is high time to strengthen the place and role of extension for national development through formulation of extension policy within the framework of broader national agricultural and rural development policies.

The formulation of extension policy should be a collaborative effort, involving all stakeholders, and should take into consideration not only technical issues but also professional development concerns with an aim to provide motivation and morale to extension workers. It should also include the operational linkages and partnerships between extension and other relevant service institutions such as related to research, marketing, environment, commodities, farm inputs, rural credit, agricultural education and training, farmer associations, and information technology.

Key Actions

- Organisation of a meeting of key stakeholders, and distribution of copies of the national policy on rural and agricultural development. The meeting should be gender-sensitive and the participants should be drawn both from the centre and provinces, and from public and non-public institutions, including small and commercial farmers' representatives. The agenda of the meeting should be:
  - Discussion on participatory drafting of an extension policy within the context of the national policy on rural and agricultural development
  - Formation of a small committee to draft the extension policy by a certain date
- Circulation of the draft policy document, when ready, among the participants of the first meeting for review and comments
- Organisation of another meeting of the stakeholders for discussion of the draft extension policy and feedback

- Finalisation of the policy document by the policy draft committee members in light of the feedback and comments received
- Approval of the extension policy through necessary government procedure
- Printing of the approved extension policy document
- Distribution of the extension policy document among the relevant public and non-public institutions related to extension work including the academic institutions where extension is taught as a subject, and representatives of farmers and other stakeholders

UPGRADING THE STATUS OF EXTENSION PROFESSION

For more than a century, agricultural extension has played a vital role in bringing about meaningful agricultural and rural development in countries like the United States of America, Australia, Japan, and the countries of Western Europe. Unfortunately, even though most extension organisations in the developing countries were established over half century ago, the profession of extension has never received the proper status it deserves, that is, in comparison to other agricultural professions.

There should be no doubt that agricultural extension is a tough and demanding profession, both physically and intellectually, yet even today, extension workers in most developing countries suffer from low salaries, meagre benefits, and negligible opportunities for development of their professional career.

A look at any major project related to research and extension, such as those financed by major donors like the World Bank, will show that most of the overseas training opportunities are grabbed by the researchers or other agricultural specialists, leaving very few, if any, for the extension staff. No surprise that few young men and women are keen to make extension their career profession after graduating from high school.

Suitable candidates will never be attracted to this profession until its service conditions are brought in line with other agricultural disciplines, and the present extension organisations are given adequate operational budget, career development opportunities, recognition, and mobility means, something essential for proper field work. Without this long overdue reform, agricultural extension in developing countries will remain a second-class profession, in spite of its demonstrated importance in bringing about rural and agricultural development in highly developed countries. The importance of extension in bringing about sustainable development and livelihoods is being recognised all over again as the fight against hunger and poverty expands worldwide.

Most of the world's poor and hungry live in rural areas. The extension workers, being relatively frequent visitors to the rural areas, if provided with appropriate service conditions, could constitute a formidable force in combating against the menaces of hunger, poverty, epidemics, and the adverse conditions resulting from natural as well as man-made disasters. However, little action has been taken so far towards improving its working conditions.

Key Actions

- Review of the conditions of agricultural extension service in comparison to other agricultural services, and policy action to rectify the situation in case any serious gaps found affecting morale of the employees.
- Meaningful involvement of key extension staff from the very beginning of any major rural and agricultural development programme of the government, such as on food security, sustainable livelihoods, desertification, alleviation of rural poverty, etc.
- Allocation of specific overseas and in-country training opportunities, scholarships for degree-seeking academic programmes and study-tours for extension discipline, and careful monitoring to ensure that they are indeed availed by the extension staff and not taken away by other disciplines.
- Frequent public recognition of extension professionals for their achievements
- A fool-proof mechanism in place to ensure that the few vehicles meant for extension field work are not taken away by non-extension government officials or politicians for personal use
- Appointment of experienced senior extension officials on policy level, decision-making positions related to rural and agricultural development, such as head of any relevant institution

PRE-SERVICE EDUCATION IN AGRICULTURAL EXTENSION

It is a well-known fact that higher education systems in most developing countries need substantial improvement. This is particularly true for the academic programmes in agricultural extension. The pre-service education in agricultural extension at under-graduate level is no more than lip service both in terms of time and content. The curricula are outdated, audio-visual aids are rarely used, suitable educational methodologies are unknown, and above all, the students of extension are given heavy doses of theory, without any exposure to real-life extension work involving rural conditions, farms and farmers. The situation at graduate level is hardly satisfactory. First of all, there are not many students who would like to select agricultural extension as their major in view of partly tough nature of extension work and partly unsatisfactory service conditions and professional development opportunities in the extension service.

There is a psychological barrier against accepting the fact that extension is not inferior to any other agricultural discipline by any means and its role is as important in nation-building programmes as of other disciplines. The poor academic programmes in extension churn out new graduates who have no technical competence and professional confidence while facing farmers, some of them for the very first time in their life. Farmers, embarrassingly for the young recruits in extension, happen to know far more of the subject these new employees are supposed to advise on. This is true that not many extension workers have aptitude of living in rural conditions, a result of both faulty selections of students for agricultural extension major as well as almost total disregard for practical training.

A worldwide movement is currently on for reforming the traditional and, by and large, obsolete agricultural and rural extension systems due to the fresh recognition that extension is bound to play a very significant role in meeting the emerging learning needs of farming communities as a result of several global developments. However, only re-orientation or in-service training of the existing extension staff will not be sufficient.

The reform and modernisation of national extension systems will remain a dream if measures are not taken for reforming pre-service education in extension. Any serious effort at reforming the national agricultural extension systems should logically start with the reform in extension education at agricultural academic institutions, which currently produce ill-prepared graduates for working in a modern extension service.

Key Actions

- Creation of a committee for review of pre-service education in agricultural and rural extension, comprising representatives from the national and provincial planning bodies, academic institutions, most relevant key government departments, non-public institutions related to rural and agricultural development, experienced and senior extension professionals, small and commercial farmers, etc., and distribution of the latest literature on extension reforms and modernisation as well as on national policy on rural and agricultural development, and terms of reference for the review
- Critical review of both under-graduate and graduate education in agricultural extension at relevant academic institutions by the review committee in order to assess the present quality and scope of the programmes against the national policy on rural and agricultural development, national development goals, and a vision of how extension services be made most useful in practical terms to face the challenge of meeting new learning needs of men and women farmers, in line with world developments. At least following aspects should be covered and recommendations for improvement made on:

  - Curricula
  - Teaching methodologies
  - Text books and other learning materials
  - Types and extent of audio-visual aids to facilitate learning
  - Training of faculty of extension in innovative pedagogy
  - Amount of practical vis-à-vis theory emphasis, especially the period to be spent in rural areas on farms with the farmers
  - Internships under real-life conditions in the field
  - Financial assistance to extension majors
  - Linkages of academic institution with the agricultural extension department and other institutions engaged in real-life extension work
  - In-country and overseas opportunities in advanced education in extension
  - Linkages with agricultural research institutions
  - Linkages with farmers' associations
  - Realisation, recognition, and emphasis on gender importance in farming and extension work
  - Gender balance in the selection of students for extension major
  - Quality and number of extension faculty and their working conditions for satisfactory teaching
  - Follow-up of the graduates after a few years in service
- Preparation of an action plan for translating the committee's recommendations into action over a specific period of time, seeking technical or financial assistance from bilateral and multi-lateral agencies, if needed.

PROMOTE PLURALISM IN EXTENSION

In early days when agricultural extension was introduced in developing countries, there was hardly any private company or meaningful non-public institution that was keen to engage in delivering extension services. That is why extension remained monopoly of the government agricultural extension departments for a long time. The situation has changed now, and one finds several public and non-public actors as well as civil society institutions, which are actively engaged in extension or extension-like activities. These actors are eager to join any formal mechanism or project through which they can make development contributions, sometimes on voluntary basis and sometimes at a fee.

Today, this is not only the government agricultural extension field worker who meets farmers to give advice, but salesmen from various commercial companies dealing in farm inputs who not only sell their products but also give advice on their use. Then, there could be several field workers belonging to charity organisations and NGOs, who are involved in extension type activities in rural areas. If extension services have been privatised in a country then individual advisors or the agents of private advisory companies keep in touch with farmers.

Some researchers who conduct research outreach activities, such as on-farm trials, also engage farmers in conversations during which quite a bit of extension advice is delivered. The same is true for university faculty and students, who come in contact with farmers for data collection purposes. Thus many countries are enjoying a pluralistic pattern of extension, but in a sporadic and haphazard manner.

Most of the time, there is no well-organised system that makes possible the active collaboration of stakeholders in both planning and implementation of extension programmes. A well-structured, pluralistic mechanism of extension planning and delivery will alleviate much of the personnel and financial pressure on the government. In addition, the farmers could benefit from a variety of human information sources, that is, in addition to the media. However, a clearly defined role of government is needed, which could comprise policy guidance, coordination, quality control, technical support to weak private organisations and NGOs, and settlement of disputes, in order to safeguard the interests of farmers.

Key Actions

— Preparation of a list of the public and non-public institutions and organisations (other than government agricultural extension department) such as private companies, NGOs, farmers' associations, rural community organisations, agricultural academic institutions, agricultural research institutes and stations, etc., in every province and/or district, which are delivering or are interested in delivering agricultural extension advice to farmers. The main aspects to be covered for each institution should be as follows:

  — The year of establishment

  — Location

  — Technical specialisations

  — Men and women professional and general services staff, with academic qualifications and field experience, whether on temporary contracts or permanent, etc.

  - Motive behind doing extension work
  - Physical facilities such as transportation, office equipment, communication facilities, audio-visual aids
  - Operational linkages with relevant institutions
  - Financial aspects such as total annual staff and programme budget, expenditure in previous year, sources of funding and their sustainability
  - Any major current programmes or those completed in recent years
  - Does the organisation work on volunteer basis or some fee is charged? If latter, details of fee.
  - Is the organisation willing to provide extension services on contract basis?
  - Does the organisation need any specialised training before it accepts contract for substantive extension work?
  - Any reports and brochures about the organisation
  - Geographical area of programme coverage
  - Type and number of people served
  - Type of extension work the organisation is interested in such as for small farmers, commercial farmers, women only, rural youth, etc.
  - Any other important information that could not be covered through above questions
- Assessment of extension and training needs of farmers in various districts and decision whether the existing government agricultural extension service is able to meet those needs. If not, the option of "outsourcing" be considered under which funding of extension and delivery of extension are considered as separate functions, and as such contracts for delivering specific extension services are given by the government to the relevant public and non-public institutions, other than the government extension department.
- Preparation of documents for outsourcing and public-private partnership, and training of the district extension staff in outsourcing the extension work to other institutions, forging public-private partnerships, and in monitoring the quality of the work done
- Periodic review and updating of the list of public and private providers of extension services

– Arrangements by the government to play its role effectively in pluralistic extension pattern in order to safeguard the interests of farmers, including responsibilities such as provision of policy guidance, ensuring coordination and quality control, provision of technical support to weak private organisations and NGOs, and settlement of disputes

PRIVATISATION OF EXTENSION SYSTEMS

In developed countries, the private sector is well developed and takes responsibility for provision of several key services, which in the developing countries are still handled by the government. The public sector is definitely dominant in less-developed countries. However, the trend towards privatisation in several developing countries is quite active these days. Agricultural extension services have been fully privatised in several European and Latin American countries.

There have been calls from certain donors and economists for privatising extension in developing countries and in fact some countries like Uganda are engaged in privatising their extension services. However, in recent years certain developments such as the disbanding of public extension services in some Latin American countries but at the same time dissatisfaction of farmers with the alternate extension services, government taking back certain components for helping farmers from the privatised advisory services in some developed countries like the United Kingdom, and uncertainty about the ultimate success of privatisation in countries where majority of farmers are subsistence, are indeed important warning signals against blind plunge into privatisation of extension in developing countries.

This should, however, not be interpreted as failure of privatisation, but as underlining the need to carefully study the feasibility of privatising extension, keeping in view the differences between the agriculture and farmers of developed countries and those of less-developed countries. The challenge is not to privatise entire extension services for all farmers, but to privatise extension where it makes sense. Under the present conditions, hundreds of millions of subsistence farmers are neither able nor willing to pay for extension advice, the quality of which in most cases does not make a convincing case for cash payment. If the extension advice is worth paying for because it enhances farmers' income, then commercial farmers, cooperatives and producer associations will be inclined to pay.

In case partial privatisation of extension is opted, proper rules, regulations and procedures will be needed both for private extension advisers and their client farmers to protect the interests and rights of all parties. In addition, the government will have to take measures for quality control of extension advice to protect farmers from being exploited. It will be prudent for developing countries to try various methods which have been tried elsewhere in the world to lessen the burden on public budget and encourage non-public parties interested in delivering extension services.

Some of the pre-requisites for proper privatisation of extension services include: the existence of competent private extension companies and individual extension advisors, strong farmers' associations with legal status which could not only effectively demand extension services but also evaluate their quality and be willing and able to pay for the advice, and a suitable government mechanism for supervision and quality control of extension services to be provided by the private sector.

Key Actions

- A survey to find out the type and number of commercial farmers, large cooperatives, and agri-businesses, which are willing and able to either hire private extension advisors or use private extension advisory services on fee basis; the larger number of willing farmers means a strong case for privatisation of extension for these particular groups and vice versa.
- A survey to find out whether a reasonable large pool of active or potential competent private extension advice providers exists or not; retired extension professionals may also be considered as potential service providers.
- A study to assess the willingness and ability of farmers' organisations, in case they exist, to pay for extension advice; the larger the number of willing farmers' associations the stronger the case for privatisation
- Assuming that the results of the surveys and the study show that the country does indeed have a strong case for fully or partially privatising agricultural extension services, the preparation of required rules and regulations and documents thereof such as contract preparation, remuneration rates, payment methods, negotiation and conflict resolution, a booklet containing a list of all private extension services providers in the area, with their qualifications and experience, technical areas of specialisation and how to contact them, government's role in supervision, quality control and conflict management, etc.
- Capacity building of individual farmers and their associations in topics such as demanding extension services, book keeping, assessing the quality of extension advice, etc.
- Retaining of an appropriate public extension force by the government for provision of free extension advice to those farmers who are either unable or not willing to pay for extension advice, and for those residing in marginal areas.

APPLY INFORMATION TECHNOLOGY TOOLS

As organisations benefit from the development of various kinds of technologies, extension organisation could also benefit from the information technology advances. The extension offices located where electricity is available could use Internet, e-mail and various advanced audio-visual equipment in their daily work. In the field, however, the benefit will be limited because of the fact that most rural areas in developing countries have neither electricity nor regular telephone facilities apart from cellular telephones.

There have been some voices calling for replacement of extension workers with information technology centres in the villages. This technology- and hardware-driven approach is not healthy, because the information and communication technology should be used for developing necessary tools that could be used by extension workers. It is well known that, in some countries, when technical recommendations were made on radio, some farmers misinterpreted the advice and damaged their crops. This was particularly true in case of recommended doses of chemical pesticides and insecticides. Later, research showed that those radio programmes for farmers which were backed up by extension workers personally, such as through radio listeners' groups, truly benefited from the extension programmes on radio.

Key Actions

It is difficult to prescribe any specific application of information technology for any specific country because conditions differ from place to place. However, the following is suggested in a generic way, which may be adapted to particular situations. As a rule of thumb, the efforts to apply information technology should be started at locations which have necessary infrastructure and pre-requisites. The other areas, obviously, will have to wait.

- A study to be conducted by a team comprising two specialists, one in electronic communication technology and the other in extension and training, in order to:
  - identify specific activities in extension work where modern information and communication technology may be realistically applied;
  - identify the extension work related locations in the country which have necessary infrastructure for installation and use of electronic technology tools;
  - prepare a plan for developing and installing necessary tools;
  - prepare a plan for training of the staff that will use those tools;
  - prepare cost estimates for the entire work
- Interactive electronic linkages may be established between extension and relevant institutions. An example is VERCON (Virtual Extension Research Communication

Network), a tool developed by FAO for facilitating operational linkages between research and extension, which remain unsatisfactory in most developing countries due to physical distances and poor conventional communication means, among other factors.

- Extension data bases may be created, containing information on last few years' prices of various commodities and projections for the near future, records of climate for the last few years and any expected unusual weather conditions in the near future, proven useful agricultural technologies, contact mailing and e-mail addresses and telephone numbers of subject-matter specialists and agricultural produce buyers, demand for grains and vegetables, etc. This database may be used by the extension workers in their work.
- A variety of attractive extension and training materials may be prepared with the help of computer, using creative techniques.
- Telecentres may be established in the areas, which are not normally covered by extension workers for reasons like remote location and inadequate staff, ensuring that periodic human follow-up is available.

STAFF ORIENTATION TO MEET MAJOR FOOD SECURITY DEMANDS

So far, in most developing countries, no real need has been felt for educating the farmers in developing countries in the issues related to globalisation, liberalisation of markets, genetic engineering and biotechnology which are sooner or later going to affect their communities. This is due to general opinion that these global level developments are of no direct concern to farmers, and in any case, are too sophisticated to be understood by them. This opinion may be right at present but it is only matter of time when these developments will start affecting the rural livelihoods.

For example, China has recently been admitted to the World Trade Organisation (WTO), and this is bound to result in significant reconsideration of existing national policies and procedures of agricultural production, agro-processing, storage, marketing, export and quality control. That is why China's national extension system must be equipped with necessary knowledge and skills in order to start educating farmers in possible effects and expectations resulting from the country's entry into WTO.

Farmers will also confront the issue of farm subsidies within the context of liberalisation of markets. In fact, it is already happening. For example, small farmers in Mali who have been growing cotton for years are in serious trouble because they are unable to compete with the very low price of the imported cotton, which the overseas farmers produced at very low cost due to heavy farm subsidies received from their respective governments.

The environmental aspects of agricultural technologies have received so far little attention both from researchers and extension workers in less-developed countries. The mandate of almost all the national agricultural extension services remains quite narrow, that is, limited to the transfer of agricultural technology. As long as a new technology can raise yields, the extension agents will fervently promote it, without any thoughts to its environmental friendliness.

During 1960s, most of the credit for the Green Revolution, besides good extension support, went to the breeding and cultivation of high yielding varieties, proper irrigation, and application of high doses of fertilisers and pesticides. That recipe for gaining high yields is, however, currently under criticism due to its unfriendliness towards conservation of environment and natural resources. This is the reason that the technologies like IPM have received favourable attention.

Rapid growth in population also needs to be given serious consideration as it offsets any gains in agricultural production. For example, between 1970 and 1995, in spite of the fact that human fertility declined in most parts of the world, Asia's population increased by 60%, i.e. an addition of over one billion people. Although a very close relationship exists between population, farm production and food security, very few extension services have paid attention to educating farmers in this very important aspect.

It may seem a bit premature at this moment, but the time is not too far when farmers will be asking extension agents the questions about biotechnology and genetic engineering. Extension services should start developing simple training materials and capacity-building programmes in these areas.

Key Actions

- As short-term measure, develop short orientation and training programmes in market liberalisation, farm-market-chain-links, quality production and market of exportable commodities and the type of strong extension support to be provided to the producers, environment sustainability, natural resources management, genetically modified crops and vegetables, and organic farming for the training of extension staff at all levels.
- Develop and use extension methodologies like APEX (Agriculture Population Environment Extension) developed by FAO, which integrate population and environment educational messages into ongoing agricultural extension programmes, after providing necessary training to the stakeholders involved.
- As a long-term measure, update various aspects of pre-service education, taking steps as explained under the guideline "Bring pre-service education in agricultural extension in line with the modernising of the national extension system".

EMPOWERMENT OF EXTENSION SERVICES

The trend to transform farming into agri-business or a commercial enterprise is visible throughout the world. The continuation of subsistence farming is being questioned on economic grounds. In many developing countries, farming has been left to old people because the new generation does not see much future in this profession. Other global developments are also hinting that the farmers cannot depend just on farm income but must diversify their income-generation habits.

There is no doubt that the farmers, if operating individually rather than collectively, will never be able to run their small farms as agri-business or commercial enterprise nor will they be able to create a strong lobby for themselves for safeguarding their interests. They must organise themselves initially in small, informal groups then gradually in special interest groups and then into larger groups, which should be registered with the government as legal entities.

A strong lobby of farmers will not only work for farmers' rights but also for demanding effective agricultural extension services from the government, especially in those countries where decentralisation or devolution has happened, thus ascertaining more budget and staff needs for the extension services. Farmers' grouping will also facilitate the application of participatory extension approaches. In summary, all these benefits will expedite the process of agricultural and rural development.

Key Actions

- Provide refresher training courses to extension workers in organising special-interest groups of men and women farmers.
- Direct extension services to organise farmers in groups and associations and have them registered with the government to enjoy legal status.
- Provide training to farmers' associations so that they could possess necessary knowledge and skills for exercising their influence in requesting specific services (such as extension advice from public and private institutions and assessing their quality when delivered, credit, farm inputs, marketing matters, etc.) as well as for converting their farms into agri-business and commercial enterprise.

OPERATIONAL LINKAGES BETWEEN EXTENSION AND RESEARCH

Extension is a service devised for the benefit of farmers, with a traditional mandate of transfer of improved agricultural technology, now being broadened in the interest of sustainable agricultural and rural development. The most important operational linkage that has been emphasized for extension since its formal introduction in the developing countries is with research. This linkage is indeed important in view of the fact that agricultural research institutes are the main source of technology for extension organisations. This is painful fact, however, that this particular linkage has been persistently and notoriously weak hence a topic of much academic and practical interest to relevant professionals for many years.

Several modalities have been tried to enhance coordination between research and extension during the last five or so decades. Some of them are mentioned below: One strategy followed by some countries was to put the two disciplines organisationally under the same administration.

The intentions were good, but unfortunately, no matter which developing country took this route, extension was grossly overshadowed and undermined by research. Research almost always ended up with the lion's share of the budget allocated. Even in major donor-financed projects, it was commonly observed that a very high percentage of overseas fellowships were taken away by research even though the project design emphasized on a fair distribution between extension and research. Another strategy to bring research and extension closer in some countries was to create small extension units in research institutes. Such units, however, did not go beyond publishing brochures and booklets on research done by those particular institutes, which was more for publicity and served as a source of pride for the researchers. In some exceptional cases, these units made research outreach efforts.

The only extension specialist running the unit found himself isolated from his professional colleagues with almost zero chances of professional growth and promotion amongst all researchers. This strategy did not work. Yet another strategy started under the World Bank-financed projects that followed the Training and Visit model of extension, was the pattern of compulsory meetings between researchers and extension workers that took place just before the start of main cropping seasons.

The purpose of the meetings was to discuss the expected field problems, identification of key messages to be passed on to "contact farmers" by the field-extension agents, and other issues related to the approaching cropping season. This modality was found to be useful, providing a productive platform for the specialists from research and extension to discuss farmers' concerns and problems. However, like the Training and Visit model, these meetings, although useful, also did not prove to be sustainable and died down with the end of particular projects. Lately, the efforts to bring research and extension closer have started taking advantage of advances in information and communication technology and virtual linkages are being established between relevant institutions.

Key Actions

- Do not repeat the old practice of putting research and extension under the same administration, as this has undermined extension. Instead, put research and extension under separate administration and take the following measures for coordination between the two disciplines:
  - Form a joint committee comprising senior officers from research and extension at the central or provincial level, depending on whether these services have been decentralised or not, for the purposes of policy advice and promotion of coordination at all levels.
  - Appoint researchers as subject-matter specialists at district level to technically backstop extension workers. If research has not been decentralised, such appointments should be for a period of about three years, during which the researchers should work under overall supervision of the district extension officer. Each researcher should spend about 30% of his time in working for his parent research institution and 70% time on supporting the extension activities in the field. The liaison between these researchers and their parent institutions is of great importance, as it not only helps the researchers to maintain technical links, but also allows their seniority to be counted, and opportunities for promotion in their respective institution will not be lost. These researchers, upon returning to their parent institutions will be considered as resource persons due to their first hand field experience leading to exchange of professional ideas with their colleagues. Working on rotation basis, these researchers should be replaced by other researchers. As most researchers do not like to be assigned at district level, special field allowance should be considered for such placements for motivation purposes.
  - Initiate a process at field level for drawing research agenda through consultation among researchers, extension workers and farmers with the objective of making the agenda field problem-oriented.
  - Making use of modern information and communication technology, establish virtual interactive linkages among research and extension institutes.
  - Make sure that there is no bias in the allocation of operational funds, establishment of service conditions, gender considerations and provision of career development opportunities to both disciplines.

REFERENCES

Leeuwis, C. 1993.. *Of computers, myths and modelling: The social construction of diversity, knowledge, information, and communication technologies in Dutch horticulture and agricultural extension.* Wageningen Studies in Sociology, 36. Wageningen: Agricultural University.

Masefield, G. B. 1950.. *A short history of agriculture in the British colonies.* Oxford: Clarendon Press.

Mook, B. T. 1982.. *The world of the Indian field administrator.* New Delhi: Vikas Publishing House.

Moris, J. 1991.. *Extension alternatives in tropical Africa.* London: Overseas Development Institute.

Plunkett, H. 1901-02.. Agricultural education for Ireland. *Department of Agriculture and Technical Instruction for Ireland, Journal,* 21., 18-41

Rivera, W. M., & Gustafson, D. J. Eds... 1991.. *Agricultural extension: Worldwide institutional evolution and forces for change.* Amsterdam and New York: Elsevier.

Smith, H. 1972.. *The Society for the Diffusion of Useful Knowledge, 1826-1846.* London: Vine Press.

Society for the Diffusion of Useful Knowledge 1827.. *Rules of the Society/or the Diffusion of Useful Knowledge.* London: William Clowes.

13

ICTs in Agricultural Extension

The concept of sustainable rural livelihoods is central to the discussion of ICTs and extension. The concept was first promoted by Chambers and Conway in *Sustainable Rural Livelihoods,* and has since formed the basis for many of the programmes of the UK Department for International Development (DFID), and is used by many overseas development agencies to orient their policies and programmes. The approach has been embraced by the UN Food and Agriculture Organisation's ICT programmes, because the link between ICTs and rural livelihoods 'facilitates the acquisition and exchange of information by the poor necessary to develop relevant livelihood strategies; improves communication within and between the institutions responsible for making decisions that affect livelihood options; and empowers poor communities to participate in decision-making processes'. The sustainable rural livelihoods approach compels us to think outside the box of agricultural extension as an approach to transferring agricultural technologies and practices to farmers.

The concept of sustainable rural livelihoods provides a way of thinking about the objectives, scope and priorities for sustainable rural development. Sustainable rural livelihoods require that farm families have the:

- Capability to perform basic functions to cope with stresses and shocks, and the ability to find and make use of livelihood opportunities
- Ability to maintain and improve livelihoods while maintaining or enhancing the assets on which livelihoods depend.

ICT interventions do not have to be specific to agriculture in order to enhance rural livelihoods or contribute to improved agricultural production. Indeed, a simple ICT intervention such as an accessible rural payphone can play a significant role in enhancing the ability of poor rural families to continue, and perhaps enhance, their contribution to national agricultural production and post-harvest activities.

The efficiency and effectiveness of agricultural extension, in the context of poverty alleviation, is directly related to the achievement of specific and measurable sustainable rural livelihoods outcomes. Agriculture—defined as the production, processing and distribution of food, fish, forest products and fibre—is the sector from which most of the rural poor derive their income. This succinct definition incorporates the important 'value-added' dimension of agriculture, through which changes in the physical state of an agricultural product, or the manner in which a product is produced and segregated, help expand the customer base and revenues available to producers, processors and marketers. There will be ICT interventions that directly relate to the agricultural sector and that contribute to sustainable rural livelihoods outcomes. There will, however, be others that are not agriculturally specific, but that have the ability to help rural families cope with stresses and shocks, and make use of livelihood opportunities.

Extension is typically seen as a service—public or private—that responds to the needs of farmers and rural people for knowledge they can use to improve their productivity, income and welfare, and to manage the natural resources on which they depend in a sustainable way. Extension brings information and new technologies to farming communities, allowing them to improve their production, income and standard of living. In the past, extension was seen primarily as a public service, institutionalised and organised by national governments. Today governments and public extension institutions are challenged to adopt diversified and pluralistic national strategies to build dialogue and collaboration among a variety of public, private, non-governmental and communitybased entities to promote extension to improve rural livelihoods. They are further challenged to adopt a 'rural extension commitment... whereby other, nonagricultural concerns are addressed—especially those relating to income-generation and the development of micro-enterprise but including health and other issues relevant to the approximately 40 percent of people in the rural sector who do not work the land'.

By focusing on how agricultural extension can use ICT to improve rural livelihoods, we need to move beyond a narrow understanding of agriculture-specific ICT applications. This change in focus causes us to recognise that:

> Any ICT intervention that improves the livelihoods of poor rural families is likely to have significant impacts (direct and indirect) on enhancing agricultural production, marketing and post-harvest activities—which, in turn, can contribute further to poverty reduction.

The challenge facing extension planners and policy-makers is how to grapple with subject matter and policy issues that are not strictly in the domain of agriculture, in order to enable agricultural extension to harness ICTs in support of sustainable rural livelihoods and poverty reduction. Extension planners and policy-makers need to be equipped with analysis and arguments to bolster the case for agricultural extension playing a broader approach in adopting and using ICTs.

Beyond Agriculture—A Broader Approach to Extension and ICTs

The rural poor are the engines of agricultural production in developing countries. Agricultural production and post-harvest activities account for the primary livelihood assets and strategies available to the rural poor. Any bottlenecks to improving the general livelihood of the rural poor—lack of health provision, disasters, lack of education, lack of infrastructure, lack of financial services, and many others—will have significant impacts on agricultural production at household, regional and national levels. ICT interventions that improve the general livelihoods of the rural poor may also yield significant agricultural development investments on the part of rural families. ICT interventions that improve the general livelihoods of poor rural families have the potential to enable those families to:

- Free up time for agricultural work through information or services that help improve family health and wellbeing
- Provide access to household capital which can be leveraged for agricultural production or post-harvest improvements
- Enable rural families to take better advantage of remittance economies, enabling family members to live elsewhere and send capital home to improve agricultural work and other livelihood activities.

Improved access to credit, educational activities, healthcare services, and improved nonagricultural rural business activities can also yield significant improvements in the agricultural sector when families are able to enhance their overall livelihood strategies. The end goal of these ICT interventions is not improved agricultural production or postharvest activities; the end goal is poverty reduction in the context of improved livelihoods, recognising the clear importance of the rural family as the hub of agricultural production in areas of poverty, and within national economies. Rethinking organisational end goals in this way could prove a challenge for agricultural extension decision-makers and their organisations—but the challenge must be faced if ICTs are to be used effectively.

EXTENSION: POTENTIAL ROLES FOR ICTS

User Demand for Effective and Appropriate Extension

Amidst the changes facing extension, there is growing recognition that farmers and members of rural communities have needs for information and appropriate learning methods that are not being met. Louder voices are being heard from rural and agricultural stakeholders, creating interest among governments and donors in supporting demand-driven extension.

Over the past 20 years, the dominant donor-supported extension approach has been public sector provision of the training and visit system (T&V). While the system is supposed to incorporate feedback from farmers, farmers are often passive recipients of didactic instruction. Messages are typically based on perceptions of farmers' needs, or on the requirements or desires of public sector agencies. Evaluation research demonstrates that T&V has not proven itself to be a system that meets the demands of users for appropriate content and learning methods.

Demand-driven extension involves a shift from public sector extension delivery to a negotiated system through which farmers and rural community members determine and identify their own needs, and have some control over the financing of extension services which are delivered by public, private, NGO or farmers' organisation providers. This approach may ensure that the delivery agencies are ultimately accountable to the client. However, 'pluralistic information flows between farmers and providers of agricultural information often experience weak linkages. These have affected communication of agricultural information, consequently causing major "bottlenecks" in national agricultural technology systems and limiting their effectiveness in contributing to agricultural development'.

In some cases ICTs are playing a very important role in supporting and facilitating emerging models of demand-driven extension. Kiplang'at reports on experiences in Kenya, South Africa, Sri Lanka and elsewhere, noting that there are some important innovations, but there is still a long way to go before a meaningful impact on extension services is seen. He notes that the trend is to integrate conventional delivery systems, such as radio and television, with new ICTs using approaches that run counter to the linear and government-owned flow of information that dominated the T&V approach. The case studies presented by Kiplang'at highlight innovative partnerships, and the increasing role of telecentres run by entrepreneurs and NGOs as conduits for a variety of communications and information flow to and from rural communities.

Extension Services and Telecommunications Policy

There is no single agricultural extension 'magic bullet' that will support demand-driven and/or decentralised/pluralistic extension, that also provides better ways to address the changes and challenges facing extension. The increasing diversity of stakeholders in extension programmes places increased importance on coordination and communication, neither of which is possible without access to basic telecommunications services such as the telephone. This widening group of 'new' actors in extension systems include:

- Farmers as both extension clients and extension providers
- Linkage, learning and knowledge management facilitators

– Private sector players
– Market players and market information providers
– NGOs, CBOs and private sector providers
– Health, education, environment and other sector players · Telecommunications players.

It thus becomes important to understand the specific actors who play (or should play) a role in establishing basic telecommunications services in rural areas. Identifying the actors can help orient projects and programmes that would lead them to effectively and collectively engage in dialogue around telecommunications policy reform. These actors include:

– Telecommunications service providers (also known as operators)
– Regulators and policy-makers
– Telecommunications policy reform advocates (most often found in, and focused on, urban service issues, and sometimes nascent in rural areas)
– Rural clients (current and potential)
– 'Last mile' entrepreneurs—phone shop operators and cybercafé/telecentre operators
– Extension managers and other professionals who deliver rural services.

The call for demand-driven extension opens the door for examination of how ICTs can be cost-effective, practical tools for facilitating and channeling farmers' demands, and addressing those demands. But when so many rural areas of developing countries lack the basic telecommunications services that support so many key ICTs, including telephone and the Internet, this examination is severely challenged. Any resulting programmes and projects remain totally dependent on the strength of national universal access. Telecommunications policy rises to the surface as a primary enabler of—or obstacle to—improving agricultural extension services.

ICT POLICY AND REGULATORY REFORM

Extension organisations play a key role in brokering between communications technologies and the client groups they serve. In this role they must be able to examine the appropriateness of various ICTs; the accessibility of ICTs in rural and remote areas; how to best to reconcile costs and benefits; and how to ensure ICT access includes a diversity of cultures, languages, social strata and age groups, and is gender-sensitive.

There is a need to temper enthusiasm about the potential use of new ICTs with a reality check on the availability of basic telecommunications services in rural and remote areas. One way to improve the situation is for agricultural extension practitioners to become more actively involved in rural telecommunications policy advocacy efforts. Even small efforts to put rural telecommunications policy on the national agenda can have notable results. Civil society advocacy efforts to improve telecommunications policy and to bridge the digital divide in El Salvador, Guatemala, Trinidad & Tobago, Canada and Australia have yielded impressive results.

However, as a recent Panos report states:

> 'In the midst of the current enthusiasm for "ICTs for development", it is often forgotten that most rural Africans do not yet even have access to telephones... In most of rural Africa, there is only one telephone for every thousand people. It is true that the number of phones in Africa has risen enormously in the past decade, especially since liberalisation [of telecommunication markets], but most of the new telephones are mobiles, and they are mostly in cities. For rural people, buying and using a mobile phone is very expensive—a single call can cost as much as half the daily wage of an agricultural worker.'

The organisations that best represent rural stakeholders—rural and agricultural organisations, and specifically agricultural extension organisations—are often absent from national policy dialogues that help create and/or shape positive policy change towards universal access. If they are not part of national policy dialogue, universal access policies, programmes and regulatory reform initiatives risk neglecting the needs of the very people the initiatives are meant to serve. There are several possible reasons for this absence:

- Rural and agricultural organisations are not directly invited to take part in national ICT policy dialogues
- The principals of these organisations have not had the opportunity to learn about effective universal access policies and regulatory instruments that, if implemented by their governments, would significantly benefit their members
- The principals of these organisations have not had the opportunity to learn about the advocacy and policy dialogue experiences, strategies and successes of their peer organisations in other countries
- ICT institutional and human capacity-building efforts are often focused on government and regulatory bodies, not on civil society actors
- The modest financing needed for capacity-building efforts that will empower rural and agricultural organisations to take part in ongoing dialogue with policy-makers is often absent from national ICT programmes.

According to the World Bank's telecommunications group:

> 'poverty reduction requires a clear understanding of the needs, priorities, demands, and pressures of the poor. Household and community surveys, and demand studies... are an

> important means of gaining this knowledge. *This is a first step in developing participatory universal access programmes,* which involve local communities in the design of such development projects through demand assessments and by allowing them to participate in the decisions about the location of particularly information access outlets.'

However, there is a great deal more that needs to be done to ensure that universal access programmes are developed using participatory processes. Empowered rural and agricultural stakeholder organisations are fundamental to ensuring that policies and programmes meet the needs of the rural poor.

The process of developing universal access telecommunications policies and programmes is an important role for regulators, and for rural and agricultural stakeholders. In order for universal access policies and programmes accurately to reflect the needs farm families, rural stakeholders themselves must be engaged in meaningful dialogue with regulators and government officials. The latter point cannot be stressed enough—and herein lies a critically important role for extension.

Too often, universal access programmes and policies are developed with the distinct absence of the rural stakeholders who those programmes and policies are meant to benefit. Their absence often results in telecoms services that do not meet their needs, and therefore generate low revenues—a problem that perpetuates the belief that rural and remote telecommunications service provision is unprofitable. Where rural stakeholders and the organisations that represent them have taken active roles in shaping universal access policies, programmes and services, the expansion of rural telecommunication services has been impressive.

ICT Interventions for Extension

The rural livelihoods approach helps draw attention to the fact that, in order to improve the efficiency and effectiveness of the agricultural sector *and* contribute to poverty alleviation, ICT project investments should be directed towards achieving the following measurable development outcomes. These outcomes will enhance the livelihoods of the people whose labour, skills, knowledge and resources make the agricultural sector function:

- *Increased farm family income* (spent on agricultural livelihood improvements, investments in small businesses, shelter, and improvements in access to basic rural infrastructure such as electricity, potable water, telecommunications and waste management)
- *Increased farm family savings* (which can be invested in livelihood strategies that directly or indirectly improve the efficiency and effectiveness of agricultural production)

- *Improved health indicators* (related to improvements in income and food security attained with relevant knowledge)
- Family investments in education and training
- *Reduced vulnerability* evidenced by indicators such as higher enrolment of young girls in school
- Reduced rural out-migration
- *Sustainable use of natural resources* evidenced by indicators such as reduced soil erosion
- *Better risk management decision-making at farm level* evidenced by indicators associated with the above outcomes.

There are many bottlenecks that can impede the achievement of these livelihood outcomes. Investments in ICT infrastructure, services and projects should acknowledge and address these obstacles:

- Challenges to the effectiveness and efficiencies of financial service organisations that serve the rural poor and agricultural producers
- Challenges to farm family access to financial services
- Legislative, policy and regulatory challenges that can impede the reach of financial services, or impede their effectiveness with the rural poor
- Lack of, or poor quality, health, education and agricultural extension training resources
- Policies, institutions or processes that impede the delivery or availability of health, education and training resources
- Government policies, institutions and programmes that may not reflect the interests or extension needs of the rural poor
- Transportation, energy, telecommunication and other sector strategies that neither address the real conditions of rural areas and ecosystem health, nor provide mechanisms for participation of the rural poor in the articulation, implementation, management and evaluation of strategies
- Rural community capacity to design, implement and manage community-based infrastructure
- Lack of private sector alternatives to public sector infrastructure provision
- Poor access to services and inputs for agricultural production, fishing and agroforestry

- Impediments to local efforts to conserve resources and enhance biodiversity through direct actions
- Government policies, institutions and programmes that might reform the supply of services, and improve management of, access to and governance of natural resources
- Lack of, or weak, environmental legislation and enforcement
- Government policies, institutions and programmes that might improve market development efforts and also increase the value of agricultural, fishery and agroforestry products
- Lack of policy and programme support for improving the internal functioning of rural and agricultural NGOs, CBOs and cooperatives
- Policies that may legally prevent or hinder the formation of rural and agricultural civil society organisations
- Lack of policy and programme support for extending the external (horizontal and vertical) links among organisations focused on enhancing sustainable rural livelihoods
- Lack of government policy and programme attention to consultation with rural and agricultural civil society organisations
- Lack of access to ICTs that supplement and/or enhance face-to-face relationships (horizontal and vertical).

ICT Project Themes for Improving Rural Livelihoods

The rural livelihoods approach yields seven most appropriate ICT project themes for improving rural livelihoods.

- Improving universal access telecommunications policies and programmes by empowering rural and agricultural stakeholder organisations so that they can participate in advocacy efforts on behalf of rural people
- Rural credit and rural financial services—improvements in access, reach and flexibility
- Louder rural and agricultural stakeholder voices yielding improved access to decision-makers to influence policies, regulations and procedures that have a direct impact on rural livelihoods
- More informed rural people and farmers who can use information to make relevant decisions about livelihood strategies, thereby reducing disaster impact (flood, disease, drought warning and mitigation), and increasing income diversification

- Improvements in efficiency and effectiveness of rural service delivery across areas of health, education, agricultural extension, training and knowledge resources
- Improved ICT planning capacities among civil society organisations—to plan, implement and integrate ICTs into their overall services
- Application of ICTs in land surveys and registration systems for more efficient recording of land titles, and registration and transfer of land holdings.

The rural livelihoods approach emphasises analysis of gender roles in the maintenance and enhancement of family livelihoods. For example, at an international scale women are twice as likely as men to be involved in agricultural activities. Women have principal roles in smallholding subsistence farming, agri-business and food processing. In most developing countries, women's income is significant to the rural household, and women are often the heads of rural households, especially where remittance economies are strong.

With regard to ICTs, there is a clear gender bias. Women significantly lag behind men in their access to, use of and production of ICTs, and in rural areas of ACP countries women are likely to be furthest removed from opportunities to harness ICTs. Priority areas for gender, ICTs and extension reflect the above development interventions, while shedding light on ways to incorporate gender:

- Training that uses local resource people, such as women entrepreneurs who are interested in, and use, ICTs
- Opportunities to develop leadership capacity within farmers' organisations and rural women's and youth groups by engaging in ICT policy debates
- Local development and testing of content oriented to women and their needs
- Training of agricultural workers, government and NGOs in gender issues and their application to the use of ICTs in agricultural extension and rural development
- Networking with other communities, countries and regions to share information and develop strategic alliances for advocacy and action.

The following case studies provide detail on some specific extension ICT project ideas and approaches which, when tied to best practices for ICT project interventions, can help improve rural livelihoods. These examples illustrate the benefits of a focus on people and their strengths; on taking time to plan initiatives by benefiting from multistakeholder input; and on a broader approach to harnessing ICTs that includes concerns about policies, institutions and processes that may either help or hinder the rural poor in their attempts to enhance livelihoods.

EXTENSION NETWORKING

Women of Uganda Network (WOUGNET)

WOUGNET is an NGO with a mission to 'promote and support the use of ICTs by women and women's organisations in Uganda, so that they can take advantage of the opportunities presented by ICTs in order to effectively address national and local problems of sustainable development.'

WOUGNET members face several challenges in developing strategies for the inclusion of Ugandan women:

- Access to the Internet is limited, especially for rural women
- Connectivity is still low across rural Uganda, and can be expensive
- Lack of electricity in rural areas
- Relevance of information to illiterate women
- Limited forms of information dissemination
- Language is an obstacle to some women

WOUGNET members are working to adapt the following creative strategies to address these challenges:

- Face-to-face networking with local municipal councils
- Face-to-face networking with local CBOs
- Collaborating with local radio stations for announcements and discussions of issues
- Better use of the postal system
- Focal point volunteers in rural areas
- Print-outs of relevant information for local noticeboards, in local languages
- Redirection of emails in print form for distribution of relevant information to other organisations
- Public dialogues that are recorded and shared online
- Participatory theatre that promotes dialogue on issues discussed online
- Posters that promote awareness of issues discussed online
- Community mobilisation for action
- Training of volunteers to train others, especially women and girls, in the use of ICTs
- Use of alternative power sources, e.g. solar

- Community-based research to ensure provision of relevant information
- Advocacy for reform of government telecommunications and ICT policies to promote universal access
- Regular meetings within member organisations for information sharing.

Solomon Islands People First Network

The Islands People First Network aims to facilitate equitable and sustainable rural development by enabling better information sharing and knowledge building among and across communities forming the Solomon Islands.

The project has established a connectivity system that permits remote locations to have Internet and email access using a computer, short-wave radio and solar power. The purpose of this telecommunications and information network is to assist the country, particularly low-income groups, in taking in charge of their own development through improved logistics, information and knowledge.

Organisationally, the Solomon Islands People First Network is established as a not-forprofit organisation within the Rural Development Volunteers Association (RDVA) through the Ministry of Provincial Government and Constituency Development. The project also plays a leading role in raising the profile of ICTs and telecommunications for development, and in assisting the government in formulating a national ICT strategy. Evaluation research on the project indicates that it is having the following impacts:

- Providing affordable, sustainable communications to the rural population (85% of the 450,000 population) where no commercial penetration has yet been possible beyond the nine provincial centres
- Using the communications network, website and partnerships to enable better networking and information sharing, especially related to rural development and peace building
- Promoting the use of ICT in development at a national level.

The People First Network is a multi-stakeholder effort that links stakeholders in government, NGOs and private telecommunications service providers. Community participation is mobilised through the formation of management committees with membership drawn from all sections of the society, including education and health workers, women's group leaders, church elders, traditional chiefs, business people and others.

The project's monitoring and evaluation efforts reveal that 'the maximum appropriation/participation occurs when the communities have identified their own needs first, and then this need has driven the demand to participate in the People First Network project'.

A key resource provided by the project is an online database of over 300 rural microprojects. The database is designed to enable project proponents, funding agencies and development support workers to leverage synergies between projects and promote collective project funding.

PestNet

PestNet is a simple, low-cost email network that helps people in the Pacific and South-East Asia obtain rapid advice and information on plant protection, including the identification and management of plant pests. Founded in 1999, the network has over 500 participating members. It links the Pacific and South-East Asian regions with plant protection specialists worldwide. The network is free to members. PestNet has members from government and NGOs, universities and the private sector, as well as farmers and students. Any organisation, group or individual can join.

Virtual Extension and Research Communication Network

The Egyptian Government established a pilot Virtual Extension and Research Communication Network (VERCON) in 2001–02, in four centres, with the support of FAO's Technical Cooperation Programme. The facility was officially launched in April 2003. The pilot was aimed at addressing the needs of small-scale Egyptian farmers through information exchange between agricultural research and extension and, indirectly, the farmers themselves. Now government researchers and extension workers based in rural villages can communicate with colleagues and experts in the other pilot locations and in the capital, in seconds rather than days.

Based on the success of the initiative in Egypt, FAO is working with several countries in the Middle East, Asia–Pacific, Africa, Latin America and Eastern Europe to develop VERCON projects. ACP countries with VERCON projects in formulation or in the pipeline include Cameroon, Kenya, Tanzania, Sudan and Uganda. The first step in establishing VERCON was to engage groups—including government ministries (agriculture, science, energy, water resources); national agricultural research centres; providers of extension services (public and private); agricultural education institutions; and farmer organisations—in a multi-stakeholder dialogue about the concept.

The VERCON project is supported by two principal partners: the Central Laboratory for Agricultural Expert Systems of the Agricultural Research Council, and the Central Administration of Agricultural Extension Services. The tools and methodologies were developed in close collaboration with the American University of Cairo. A multi-stakeholder VERCON Steering Committee has helped to develop links with other ministries and NGOs active at the community level. Following the completion of the pilot phase, training, computers and Internet access have been provided to more than 50 government institutions. With funding provided by the Italian Government as well as contributions from the Egyptian Government, FAO is helping to expand the initiative to include agricultural training centres and the private sector across the country. As a result, the VERCON pilot is poised to become more than a network for public sector extension and research, and may emerge as a system that enables village-level stakeholders to communicate with national and regional counterparts, and to share and access information across the Arabic-speaking world.

VERCON's innovation is in its capability to achieve effective linkages by connecting geographically dispersed people, and to enhance two-way communication, managing large volumes of data, and rapidly collecting, processing and dispersing information in a variety of forms. The human component is a network (e.g. staff of research and extension institutions, faculties of agricultural education, NGO workers and agricultural producers) committed to strengthening collaboration, communicating, sharing information and supporting improved agricultural production.

RURAL ICT TOOLKIT

The African Connection Program draws attention to telecommunications and ICT strategies and activities that will assist poorer African countries. The *Rural ICT Toolkit for Africa* brings together a compilation of essentials, best practices and checklists for planning, financing and implementing telecommunications and ICT projects in the rural African context. The toolkit includes:

- Glossary of ICT elements, technical and non-technical terms and project types
- Principles of the African Connection Program's rural ICT development programme
- Best practice for rural ICT policy
- Country market situation—what kind of rural ICT projects are appropriate to, and possible in, a country's stage of policy, infrastructure and market development?
- Basic introductions to technology niches and applications
- Market dynamics and demand analysis
- Business planning
- Establishing the socio-economic impact
- Funding principles and processes
- Monitoring and evaluation.

The toolkit is a key resource for the African Connection Program's funding and promotional activities, which include the following:

- Rural public calling offices (PCOs)
- Regional 'next-level' Internet points of presence to encourage local Internet access in rural district centres. At these points, intermediate agencies with active rural programmes, social infrastructure institutions, vanguard schools, health institutions, and small and medium enterprises will be able to gain access to the Internet
- Networks for rural intermediate agencies to enable and support the implementation of computer networks and other ICT needs. These agencies include NGOs, microcredit, local government and social infrastructure agencies (education, health, development, etc.) that have direct activities in rural areas
- Small-scale telecentres—instead of financing and owning the installation and infrastructure for large multi-purpose community telecentres, the programme will operate through small business matching grants or loans and provision of business plan assistance. The aim is to facilitate the bottom-up expansion of already existing and successful PCO businesses into the next stage of ICT service provision to become info kiosks or mini telecentres; and to maximise existing potential for privately owned and operated telecentres to expand their activities into training, capacity-building and rural community outreach through partnership with donor agencies underwriting demand for these services
- ICT content—support the development of relevant and appropriate content to promote ICT diffusion in rural areas
- Policy and regulatory support—promote and support policy and regulatory initiatives that can act as a catalyst or provide leverage to achieve the above.

PARTICIPATORY 3D MAPPING

The Philippine Association For Intercultural Development (PAFID) is an NGO that assists indigenous communities to regain and secure ancestral domains (PAFID, 2004). In Mindanao the Government of the Philippines has recognised some ancestral domains, but the boundaries have not yet been properly identified and mapped. This situation has caused frustration in the process of settling ancestral domain claim applications. In partnership with the International Fund for Agricultural Development (IFAD), PAFID is engaging in participatory community mapping to help bring about full recognition of the rights of indigenous communities over their ancestral domains.

PAFID is using an innovative participatory 3D mapping process that marries sophisticated ICTs with common household tools. The process begins with community consultations to access information about ancestral territory, and needs and obligations with regard to the land. Participants draw rough maps that identify territorial boundaries, land-use histories and patterns, and significant geographic and ecological features. The locations of these boundaries and features are verified by global positioning system (GPS) ground surveys, resulting in refinements of the rough maps. Using cardboard, paint, yarn, push-pins and other simple modelling tools, participants sculpt topographic maps to create 3D geographic models that illustrate elevations, contours, natural resources, land cover, settlements, infrastructure and administrative boundaries.

The 3D geographic models, created with cardboard and other household tools, are digitally photographed and the photographs are joined into a geographic information system (GIS). The images are corrected using GPS ground survey data and combined to produce 2D thematic maps of land cover, natural resources and other significant features.

PAFID and its community partners have mapped over a million hectares of ancestral domains, and the resulting maps are now accepted by the government as proof of claims for legal recognition of ancestral land rights. The success of the participatory 3D mapping approach is based on its ability to generate an ongoing and mutually beneficial political dialogue between indigenous communities and government authorities.

AREAS FOR ACTION

The following recommendations flow from the Centre for Agricultural and Rural Cooperation (CTA). Each represents an area for action in which extension can or should play a more effective role.

Extension Advocacy for Basic Voice Telephony for Improving Rural Livelihoods

Within agricultural activities, telephones are commonly used to gather information about market prices, market trends, and availability and cost of inputs, and for making market transportation and sales arrangements. Telephones enable farmers to take advantage of market information, increase profits, and reduce productive expenses. Telephones are also a critical tool to reduce the risks involved with remittance transfers, and to enable farm family members to obtain accurate information about foreign currency exchange rates.

The telephone can also be an important tool in financial processes and transactions, and the example of their use in remittance payment systems is indicative. Where available, the use of telephones to facilitate remittances and reduce risks is a significant element in the livelihood strategies of farm families. At the micro-level, remittances tend to be used for daily household expenses such as food, clothing and healthcare.

Remittances are thus an important factor in meeting household subsistence needs, and can make up a significant portion of household income. Remittance funds are also spent on capital items, including building or improving housing, buying cattle or land, and buying consumer goods such as portable radio/tape/CD players and televisions. Once subsistence needs are met, remittances tend to be used for productive investments in the agricultural sector, or in other family business activities, or for savings. Extension can play a key role as an advocate for ICT policy and regulatory reform focused on basic voice telephony.

Rural Telecentres

Based on the author's rural telecentre work in Thailand and Sri Lanka, and on analyses of telecentre initiatives in Africa, the Caribbean and Latin America, it is easy to conclude that donor-funded telecentres, while able to provide some temporarily useful services, are generally unsustainable. The telecentres that are sustainable are generally operated by entrepreneurs or entrepreneurially oriented NGOs or CBOs, and focus on provision of basic services such as voice telephony, basic email access and computer training. Any programme that provides core capital and/or operational funding for rural telecentres should be approached very cautiously. However, there are important roles for entrepreneurially driven telecentres as:

- One tool in a universal access policy or programme (especially one that has involved rural and agricultural stakeholders in its design)
- A focus for micro-credit loans—entrepreneurs who operate rural telecentres have significant difficulty in accessing credit
- A rural location for practical ICT access to financial services—an opportunity to provide 'win–win' service offerings between the financial sector and telecentre operators; or for financial sector services to expand to include basic ICT access; or for telecoms operators to expand their services to include financial services
- Excellent vehicles to facilitate louder rural and agricultural stakeholder voices
- Excellent vehicles for improving rural and agricultural stakeholder access to, and dissemination of, training and knowledge resources.

Support for telecentres should emerge as a by-product of work designed to achieve rural livelihoods outcomes, and which is tied to parallel, multi-stakeholder objectives and activities. 'Silo' telecentre initiatives, developed in the absence of a focus on poverty alleviation or sustainable rural livelihoods, will be fraught with sustainability problems.

The e-Sri Lanka initiative funded by the World Bank illustrates a positive break in the tendency for donors to fund rural telecentres directly. This programme is rolling out more than 200 rural telecentres across the country, using a competitive approach to selecting entrepreneurs who will operate the equivalent of village telecentre franchises. The programme has been developed from the bottom up, with extensive consultation with and participation of rural community leaders from across the country. The approach to this initiative is based on a pilot programme established in Thailand to determine the most appropriate telecentre operational model and programme for government support. In the case of Thailand, the most appropriate and financially viable model was a telecentre owned and operated by village entrepreneurs.

Extension needs to work more closely with rural telecentre operators and the programmes that support them. Telecentres are a basic conduit for a huge variety of communication and information sharing, a conduit that is seriously under-utilised.

Planning for Effective ICT Application

ICTs have unique features that provide opportunities to harness them in ways that are different from how the traditional media have been used for development. According to Michiels and Van Crowder, in comparison with traditional media ICTs can achieve the following:

- Offer opportunities for two-way and horizontal communication, and for opening up new communication channels for rural communities and the intermediaries and development organisations that support them. Once mastered, they potentially allow every user to be a sender, receiver, 'narrowcaster' and broadcaster
- Support bottom-up articulation of development needs and perceptions, and facilitate the merging of global and local knowledge and information
- Support, create and strengthen interactive and collaborative networks that enable information to flow to and from rural communities; facilitate dialogue between communities, intermediaries and development organisations; foster coordination of national and local development efforts; and overcome physical barriers to knowledge and information sharing. ICTs can also enhance the capacity of grassroots organisations to make their voices heard. This is especially true of ICT projects that are managed by local communities, such as community-owned media and community telecentres
- Support policy and advocacy by meeting the information needs of elected officials, decision-makers, interest groups and grassroots advocacy organisations. They can be activated for social networking and mobilisation, to build up public awareness around development issues, and for upward pressure on policy decisions
- Help build consensus through the provision of information on government programmes, policies, decisions and issues to advocates. Many governments are

making such information available online. On the other hand, opponents can also seize the same tools for Internet campaigns to support their own agendas. Such online 'checks and balances' of political agendas potentially can contribute to political debate and democratic processes

– Enhance partnership with the media. They are particularly relevant for community media that have limited human and financial resources.

Michiels and Van Crowder challenge us to harness new ICTs, especially for rural and agricultural development:

> 'Until people start to capitalise on the various experiences in experimenting with, adapting, and "transforming" the new technologies, community-based ICT projects may not be seen to offer any real or direct benefits. Benefits are even more difficult to reap given that ICTs are found mainly in urban centres and thus are largely out of reach of people living in rural areas.'

Good planning is imperative. Developing agricultural extension applications of ICTs often involves gaining institutional or organisational support. Gaining such support is not without its challenges. It is critical to ask whether the initiative you are developing dovetails with your organisation's core business. If it does not, your prospects of success are limited. The individuals involved within an organisation must see direct benefits in applying their time and resources to a communication for development effort. This is especially true if they are being asked to support time-consuming multi-stakeholder planning. Organisations that already have a track record of supporting such planning will be more predisposed to support the planning required for communication for development efforts.

Multi-stakeholder planning is very important in ICT projects. Given the extensive range of networking enabled by new media, dialogue with like-minded agencies and with agencies working to achieve similar goals or working in the same contexts, can yield many benefits. Benefits of multi-stakeholder planning include shared capital and/or operating costs, optimised strategies, economies of scale, and improved polices and programmes.

The rural livelihoods approach provides some specific programming directions based on appropriate outputs for ICT interventions that will enhance sustainable rural livelihoods and the agricultural sector. These recommendations, combined with best practices for project planning and implementation, provide an excellent entry point for agricultural extension planners and decision-makers. The discussion below draws from the latest literature in the field of ICT for development.

Curtain, in a paper that draws on a database of 100 ICT-based projects in developing countries, provides some important orientation:

- Start with a broad definition of ICTs that includes both old and new forms of available technologies, and recognise that newspapers, radio and TV offer considerable unrealised potential, including potential that can be realised through the marriage of old and new ICTs
- Start with an assessment of readiness to use ICTs—in this case at national and rural levels, and with prospective organisations at the appropriate level of intervention
- Determine whether you are proposing an ICT-driven project or an 'ICT-supported mainstream project'; Curtain (2004) points out that the planning questions you need to address for each approach are very different.

Based on FAO's experience with the Virtual Extension and Research Communication Network and FarmNet initiatives, and IFAD's experience with its Electronic Networking for Rural Asia–Pacific project, the following recommendations supplement Curtain's:

- Be very cautious about supporting ICT-driven projects that are heavily focused on technical solutions to human problems. These 'magic bullet' solutions are prone to failure because they tend to neglect to deal with governance, financing and other sustainability issues
- Begin by addressing the involvement of the full community of ICT users, not just half of the community: ensure women and girls are at the decision-making table from day one
- The user community must have a financial commitment/investment in the project and, ideally, has already identified appropriate uses for ICTs and is making financial commitments in this direction
- Conduct communication and information-sharing readiness assessments that focus on human and organisational dimensions, as opposed to technical dimensions. In other words, find out how well people are currently communicating, and why. Organisations that are predisposed to open and transparent communication are more likely to adopt and adapt ICTs to gain further advantages. Those organisations that do not like to disclose or share information are not likely to gain the full value of ICTs, and will probably not perform well in multi-stakeholder initiatives. Rigid managerial control over ICT access is a common phenomenon
- Focus initial project management attention on sensitising organisational decisionmakers to the importance of their strategic direction-setting, project governance, and the likely need to modify organisational policies and procedures to gain full value from ICTs
- ICT projects are multi-stakeholder in nature and must include multi-stakeholder governance, management, and monitoring and evaluation. This parallels a

fundamental IT planning principle—participation of the user community in design and management of the IT solutions that have an impact on their work

- Unless a project has a clear local 'champion' organisation or individual, do not provide support
- Trusted partners and major influencers of the farm/rural family often have better access to IT, are a smaller, more cost-effective target group, and can have greater economic impacts then ICT projects aimed directly at the farm/rural family
- ICT projects inserted into the agriculture supply chain must demonstrate tangible economic benefit for all stakeholders involved.

Multi-stakeholder planning

Because new ICTs are generally dependent on access to telecommunications services, it is important to attempt to involve telecommunications service providers, software vendors and equipment vendors in multi-stakeholder planning for ICT projects. Given the challenges of supporting ICT projects in rural areas, any 'win–win' collaboration between development agencies and technical service providers that expand rural telecommunication services will yield multiple benefits to rural communities. Organisations must also be oriented towards open and transparent communication. People need to be able to freely, openly and creatively express and share their personal and professional goals, in ways that allow all stakeholders to learn about one another's goals. Multi-stakeholder planning also involves internal participants and external stakeholders in identifying, refining and ranking goals for improving relationships, partnerships and networks. Organisations with experience in multi-stakeholder planning will have an easier time supporting communication for development efforts than those that do not have this experience.

Measurement frameworks are critical to fostering and determining the success of any communication for development effort. Measurement frameworks must be relevant and meaningful to stakeholders. To this end, frameworks must be developed by the participants, and the tracking of measurement indicators must also be done with and by participants. This will help ensure that participants take responsibility for the evolution of tools and processes that work. Again, organisations with experience in using participatory techniques to develop measurement frameworks will have an easier time supporting communication for development efforts than those that do not have this experience.

It is also important to recognise that organisational incentives to use communication for development approaches are often weak. Other criteria (e.g. securing funding, rewarding supports) may be more important in determining the success of a manager or a policymaker than the outcomes of a specific communication for development project. Organisations that tend to reward managers and policy-makers for the development outcomes of their projects will be more inclined to support communication for development initiatives.

Finally, organisations may not always provide smooth budgetary and technical support for some of the basic features of communication for development efforts that involve new ICTs. Support for access to telecommunications services (e.g. monthly phone bills and Internet service provider bills), and support for software and hardware, are absolutely necessary. There is a need for budgetary planning awareness and integration of initiatives within budgetary cycles and strategic planning. Where organisations lack the dimensions described above, champions of communication for development initiatives involving new ICTs will probably have to engage in internal advocacy efforts to improve specific situations. Armed with good research that ties an initiative to an organisation's goals and objectives, champions will have a much better chance of succeeding in accessing the support they require.

REFERENCES

Batchelor, S. and Scott, N. 2001. *The Role of ICTs in the Development of Sustainable Livelihoods: A Set of Tables.* London: Sustainable Initiatives.

Greenridge, C. 2003. Welcome Address: ICTs Transforming Agricultural Extension? Presentation to CTA's Sixth Consultative Expert Meeting of its Observatory ICTs. Wageningen, the Netherlands: CTA.

Irigoin, M., Whitacre, P., Faulkner, D. and Coe, G., eds. 2002. *Mapping Competencies for* Communication for Development and Social Change: Turning Knowledge, Skills, and *Attitudes Into Action.* Washington, DC: Academy for Educational Development.

Michiels, S. and Van Crowder, L. 2001. *Discovering the "Magic Box": Local Appropriation of Information and Communication Technologies (ICTs).* Rome: Food and Agriculture Organization.

Nicol, C., ed. 2003. *ICT Policy: A Beginner's Handbook.* San Francisco, CA, USA: Association for Progressive Communications.

Richardson, D. 1996. *The Internet and Rural Development: Recommendations for Strategy and Activity.* Rome: Food and Agriculture Organization.

WOUGNET. 2003. *WOUGNET Meeting – Information Sharing on the Application of ICTs, 15 January 2003.* Kampala: Women of Uganda Network.

14

Management of Extension Programmes

Managing people effectively in extension programmes is a skill that requires constant planning and development. An extension programme manager can be defined as the person who is vested with formal authority over an organisation or one of its sub units. He or she has status that leads to various interpersonal relations, and from this comes access to information. Information, in turn, enables the manager to devise strategies, make decisions, and implement action.

Management is concerned with the optimum attainment of organisational goals and objectives with and through other people. Extension management organisations are characterised by many strategies, wide spans of control, democracy, and autonomy. Their management practices cannot be reduced to one standard set of operating guidelines that will work for all organisations continually. However, all managers of professional organisations face the same challenge: to manage one's time, objectives, and resources in order to accomplish tasks and implement ideas.

Managers of extension programmes are painfully aware of the need for revision and development of the new skill sets held by today's high performers. If change is not handled correctly, it can be more devastating then ever before. High performers reflect, discover, assess, and act. They know that a new focus on connecting the heads, hearts, and hands of people in their organisation is necessary.

Astute managers know what needs to be done but struggle with how to do it. Quite often they prefer to consider themselves as teachers or communicators rather than managers. This results in under-utilisation of the increasing amount of literature on management theory and practice. The root of the problem is implementation. They must learn how to motivate others and build an efficient team.

More formally defined, management is the process by which people, technology, job tasks, and other resources are combined and coordinated so as to effectively achieve organisational objectives. A process or function is a group of related activities contributing to a larger action. Management functions are based on a common philosophy and approach. They centre around the following:

1. Developing and clarifying mission, policies, and objectives of the agency or organisation
2. Establishing formal and informal organisational structures as a means of delegating authority and sharing responsibilities
3. Setting priorities and reviewing and revising objectives in terms of changing demands
4. Maintaining effective communications within the working group, with other groups, and with the larger community
5. Selecting, motivating, training, and appraising staff
6. Securing funds and managing budgets; evaluating accomplishments and
7. Being accountable to staff, the larger enterprise, and to the community at large.

The management functions can be categorised by using the acronym POSDCORB:

- *Planning:* outlining philosophy, policy, objectives, and resultant things to be accomplished, and the techniques for accomplishment
- *Organising:* establishing structures and systems through which activities are arranged, defined, and coordinated in terms of some specific objectives
- *Staffing:* fulfilling the personnel function, which includes selecting and training staff and maintaining favourable work conditions
- *Directing:* making decisions, embodying decisions in instructions, and serving as the leader of the enterprise
- *Coordinating:* interrelating the various parts of the work
- *Reporting:* keeping those to whom you are responsible, including both staff and public, informed
- *Budgeting:* making financial plans, maintaining accounting and management control of revenue, and keeping costs in line with objectives

PLANNING

Planning is the key management function of any extension worker. It is the process of determining in advance what should be accomplished, when, by whom, how, and at what cost. Regardless of whether it is planning long-term program priorities or planning a two-hour meeting, the planning aspect of management is the major contributor to success and productivity. Stated simply, "If you don't know where you are going, then you won't know when you have arrived!" Planning is the process of determining the organisation's goals and objectives and making the provisions for their achievement. It involves choosing a course of action from available alternatives.

Planning is the process of determining organisational aims, developing premises about the current environment, selecting the course of action, initiating activities required to transform plans into action, and evaluating the outcome. The types of planning that managers engage in will depend on their level in the organisation and on the size and type of the organisation.

Generally there are four major types of planning exercises: strategic, tactical, contingency, and managerial. Strategic planning involves determining organisational goals and how to achieve them. This usually occurs at the top management level. Tactical planning is concerned with implementing the strategic plans and involves middle and lower management. Contingency planning anticipates possible problems or changes that may occur in the future and prepares to deal with them effectively as they arise. Managerial planning is usually considered as microlevel planning. It helps in combining resources to fulfil the overall objectives of the extension organisation.

A needs assessment may initiate a need for developing a plan. The planning process begins with the creation of a philosophy that consists of statements describing the values, beliefs, and attitudes of the organisation. Its mission statement is a proclamation of its purpose or reason for being. After the philosophy and mission statements have been established, various goals and objectives are defined. Goals are usually general statements that project what is to be accomplished in the future.

An objective is a concrete statement describing a specific action. Policies are predetermined guides to decision making; they establish boundaries or limits within which action may be taken. Managers are related to policy formation in two ways. First, they play a crucial role in implementing organisational policies that have been established by higher management. Second, they create policies within their departments as guides for their own work groups. Procedures outline the series of steps to be followed when carrying out a designed policy or taking a particular course of action. Rules are used to provide final and definite instruction. Usually they are inflexible. Planning is designing the future, anticipating problems, and imagining success. In short, planning is essential for anyone who wants to survive.

The functions of organising, leading, staffing, and budgeting are means of carrying out the decisions of planning. Everyone is a planner—a planner of meals, of work time, Of vacations, of families. Formal planning, however, distinguishes managers from non-managers, effective managers from ineffective managers. Formal planning forces managers to think of the future, to set priorities, to encourage creativity, to articulate clear objectives, and to forecast the future in terms of anticipated problems and political realities.

Long-range Planning

Long-range planning is vitally important in that it focuses attention on crucial future issues which are vitally important to the organisation. It involves studying societal trends and issues, surveying current and anticipated learners' needs, and being aware of long-term research directions and changes in technology. Many extension workers may think that such management is beyond their level of authority, control, or involvement. They may feel that such management is the prerogative of the director, the deputy minister, or the president. However, while senior levels of management must be involved, those who implement the objectives resulting from long-range planning should also be involved.

Strategic Planning

Strategic planning has been defined as that which has to do with determining the basic objectives of an organisation and allocating resources to their accomplishment. A strategy determines the direction in which an organisation needs to move to fulfil its mission. A strategic plan acts as a road map for carrying out the strategy and achieving long-term results. Occasionally a large gap exists between the strategic plan and real results. To boost organisational performance, people must be a key part of the strategy. A stronger, more capable and efficient organisation can arise by defining how its members can support the overall strategy.

Strategic planning is different from long-term planning. Long-range planning builds on current goals and practices and proposes modifications for the future. Strategic planning, however, considers changes or anticipated changes in the environment that suggest more radical moves away from current practices. When doing strategic planning, the organisation should emphasize team planning. By involving those affected by the plan, the manger builds an organisation wide understanding and commitment to the strategic plan. The elements of strategic plans include:

- Organisation mission statement—What
- Strategic analysis—Why
- Strategic formulation—Where
- Long-term objectives implementation—When and How
- Operational plans—When and How

Changes are essential to better position the extension organisation and focus on client needs and moving forward in rural development and sustainability programmes. The strength and resilience of the traditional rural and farm population and the trend towards a decentralised society with more and more urbanites moving to the country suggest that successful rural communities will depend on people's ability to change, to adapt, and to work toward a better future. In the 1990s, facilitating farmer participation is a major extension activity. Reorganisation provides a framework for longer-term commitment to rural development.

Organisations and sub units are being encouraged to put work teams in place to ensure that each sector integrates staff and services into a cohesive, focused business unit. Consultation and participation are believed to be essential for the successful development and implementation of organisational goals and objectives. Each work team is asked to develop an effective process for discussion of major challenges and opportunities facing the organisation, if possible, over the next decade. Updated strategic plans are then developed. These plans form the framework for focusing organisational resources on the most strategic areas by using a staged approach. Updated plans are then implemented by work teams at all levels of management. Work-team objectives include:

1. Involving all levels of staff in consultation
2. Designing and implementing a process to develop-goals and objectives for the organisation and unit; a strategic process for the next five to ten years
3. Defining and clarifying organisational structures and identifying functions, customers, and service delivery models
4. Identifying changes and staged approaches needed to move from the current situation to what will be required over the next three to five years
5. Identifying and recommending priorities for policy and programme development
6. Incorporating goals for expenditure reduction, service quality improvement, workforce management, accountability, technology, and business process improvement
7. Stating the start date and first report date

Managerial Planning

If long-range planning can be linked to "macro," then managerial planning can be linked to "micro." Managerial planning is the implementing of the strategic plan; it is the combining of resources to fulfil the overall objectives and missions of the organisation. Managerial planning focuses on the activity of a specific unit and involves what needs to be done, by whom, when, and at what cost. The strategic planning process serves as an umbrella over the management planning process which deals with the following:

1. Establishing individual goals and objectives
2. Forecasting results and potential problems
3. Developing alternatives, selecting alternatives, and setting priorities
4. Developing associated budgets
5. Establishing personnel inputs
6. Establishing specific policies related to the unit
7. Allocating physical resources
8. Appraising how the management unit has succeeded in meeting its goals and objectives

DECISION MAKING

Closely related to both strategic and managerial planning is the process of decision making. Decisions need to be made wisely under varying circumstances with different amounts of knowledge about alternatives and consequences. Decisions are concerned with the future and may be made under conditions of certainty, conditions of risk, or conditions of uncertainty. Under conditions of certainty, managers have sufficient or complete information and know exactly what the outcome of their decision will be. Managers are usually faced with a less certain environment. They may, however, know the probabilities and possible outcomes of their decisions, even though they cannot guarantee which particular outcome will actually occur. In such cases, there is a risk associated with the decision and there is a possibility of an adverse outcome.

Most managerial decisions involve varying degrees of uncertainty. This is a key part of a manager's activities. They must decide what goals or opportunities will be pursued, what resources are available, and who will perform designated tasks. Decision making, in this context, is more than making up your mind. It consists of several steps:

- Step 1: Identifying and defining the problem
- Step 2: Developing various alternatives
- Step 3: Evaluating alternatives
- Step 4: Selecting an alternative
- Step 5: Implementing the alternative
- Step 6: Evaluating both the actual decision and the decision-making process

Managers have to vary their approach to decision making, depending on the particular situation and person or people involved. The above steps are not a fixed procedure, however; they are more a process, a system, or an approach. They force one to realise that there are usually alternatives and that one should not be pressured into making a quick decision without looking at the implications. This is especially true in the case of nonprogrammed decisions as contrasted to programmed decisions. One of the most difficult steps in the decision-making process is to develop the various alternatives. For example, if one is involved in planning a workshop, one of the most crucial decisions is the time, format, and location of the workshop. In this case, one's experience as well as one's understanding of the clientele group greatly influence the selecting of alternatives. Often decision trees can help a manager make a series of decisions involving uncertain events.

A decision tree is a device that displays graphically the various actions that a manager can take and shows how those actions will relate to the attainment of future events. Each branch represents an alternative course of action. To make a decision tree it is necessary to:

(1) identify the points of decision and alternatives available at each point,

(2) identify the points of uncertainty and the type or range of alternative outcomes at each point,

(3) estimate the probabilities of different events or results of action and the costs and gains associated with these actions, and

(4) analyse the alternative values to choose the next course of action.

In extension, the decision-making process is often a group process. Consequently, the manager must apply principles of democratic decision making since those involved in the decision-making process will feel an interest in the results of the process. In such a case, the manager becomes more of a coach, knowing the mission, objectives, and the process, but involving those players who must help in actually achieving the goal. The effective manager thus perceives himself or herself as the controller of the decision-making process rather than as the maker of the organisation's or agency's decision.

ORGANISING

Once strategic planning and management planning are implemented, organising to get the job done is next. Organising is the process of establishing formal relationships among people and resources in order to reach specific goals and objectives. The process, according to Marshall, is based on five organising principles: unity of command, span of control, delegation of authority, homogeneous assignment, and flexibility. The organising process involves five steps: determining the tasks to be accomplished, subdividing major tasks into individual activities, assigning specific activities to individuals, providing necessary resources, and designing the organisational relationships needed. In any organising effort, managers must choose an appropriate structure. Organisational structure is represented primarily by an organisational chart. It specifies who is to do what and how it will be accomplished. The organising stage provides directions for achieving the planning results. There are several aspects to organising—time, structures, chain of command, degree of centralisation, and role specification.

Time Management

Managers must decide what to do, when, where, how, and by or with whom. Time management is the process of monitoring, analysing, and revising your plan until it works. Effective planning is a skill that takes time to acquire. It is difficult to implement because you have no one but yourself to monitor how effectively you are using your time. Everyone has the same amount of time—168 hours per week. How that time is managed is up to the discretion of each person. One extension agent joked that he was so busy taking time management courses, he had little time left to manage. Effective time management involves philosophy and common sense. Time is not a renewable resource—once it is gone, it is gone forever. To function effectively, managers have to be able to prioritise and replace less important tasks with more important ones. Most of us work for pay for only 1,800 hours per year. Effective and efficient time management encourages us to achieve and be productive while developing good employee relations.

Goals should be specific, measurable, attainable, realistic, and timely (SMART). Once the goals are known, it is important to think about how they can be achieved. Effective time managers facilitate planning by listing tasks that require their attention, estimating the amount of time each task will take to complete, and prioritising them—deciding what tasks are most important to do first and numbering them in rank order. It is essential to know what is crucial and what is not.

Some activities have relatively low levels of importance in completing a given task. By planning ahead, managers can decide what to do and take the time to come up with ideas on how to do it. They can make their own list of steps to eliminate or reduce time wasters. Maintaining a daily "To Do" list with priorities attached and maintaining a daily, weekly, monthly, and yearly diary is helpful. Managers should analyse their daily activities to see which are directed toward results and which are simply activities. They could learn how to manage meetings more effectively since considerable management time seems to be wasted in nondirectional formal meetings. One of the methods that helps allocate time according to priorities is Pareto's Law or the 80/20 rule: if all items are arranged in order of value, 80 per cent of the value will come from 20 per cent of the items. For example, 80 per cent of the complaining in your department is likely to be done by 20 per cent of your staff. Four suggestions for better time management are:

(1) never handle the same piece of paper twice;

(2) learn how to say "no" without feeling guilty about requests that do not contribute to the achievement of your goals;

(3) when a visitor drops in to your office, stand up while you have your discussion to ensure that only a brief period of time will be consumed by the visitor's interruption; and

(4) avoid being a slave to the telephone.

By managing time well, managers are better able to solve problems quickly, make decisions, avoid frustration, keep from getting bogged down in day-to-day tasks, handle crises, work on their goals and priorities, and manage stress. Guidelines for scheduling time include:

1. Always put your schedule in writing.
2. Focus on the objectives you are trying to accomplish.
3. Continually review objectives, priorities, and scheduled actions to keep on track.
4. Schedule around key events and actions.
5. Get a productive start by scheduling early-day actions.
6. Group related items and actions whenever possible.
7. Do not hesitate to take large time blocks for important tasks.
8. Be sure to allow enough time for each task, but not too much time.
9. Build in flexibility for unexpected events.
10. Include some thinking time for yourself.
11. Consider how to make waiting and travel time useful or otherwise productive.
12. Try to match your work cycles to your body cycles.
13. Learn to control your unscheduled action impulses.
14. Prepare tomorrow's schedule before you get to the office in the morning.

Structures

Working productively and developing feelings of cooperation and effectiveness are related to having the right people doing the right jobs. Structure, then, can be defined as a system of interrelated jobs, groups of jobs, and authority. There is no standard organisational structure, but most organisations and agencies follow the "Christmas Tree" system with the star (e.g., president, minister) at the top, smaller branches at management levels, and bigger branches at the production levels. Some would claim that the lower branches support the upper branches, but as in the tree, the branches are supported by a single trunk, which can be thought of as the organisational mission and objectives. Each part of the tree has its specific function. When all parts work together, the system survives, functions productively, has balance, and is a pleasure to see! There are four primary elements in designing an organisational structure:

1. *Job specifications*—what each division/office/unit is responsible for
2. *Departmentalisation*—the grouping of jobs and responsibilities in common sectors with the objective of achieving coordination
3. *Span of control*—a definition of how many job roles should be in each unit and which roles require coordination by a unit manager
4. *Delegation of authority*—assigning the right to make decisions without having to obtain approval from a supervisor

The resulting organisational structure will vary according to these four elements. An organisation with decentralised authority and very heterogeneous departments will appear very different from one with centralised authority and a very homogeneous product.

Chain of Command

Once an organisation starts delegating authority, then there is automatically a chain of command, "the formal channel which specifies the authority, responsibility and communication relationships from top to bottom in an organisation". Thus authority flows from presidents to vice-presidents to divisional managers, from ministers to deputies to directors, from principals to vice-principals to deans, etc. In complex organisations, there may be bridges from one level to another and there will be complex procedures for maintaining the chain of command. Adult and extension educators, if working for an organisation or agency, will be part of a structure and part of the chain of command.

One cannot often make major changes in these two elements; it is wise, however, to be very aware of the organisational structure and chain of command if you wish to accomplish things efficiently. Centralised organisations are those in which the key authority and decision-making role is focused on one or a very few individuals. Where authority is distributed among many managers, then one can see a decentralised structure. As the organisation's various roles become more diverse in terms of programme, product, or geographical location, one can see a more decentralised organisational structure with authority being delegated to those who are closest to the action. Centralisation refers to authority, whereas centrality refers to the proximity to the organisation's stated mandate and objectives. One could have a very decentralised organisation with each unit being responsible for programmes, staffing, and budget, and yet be very close to the main mission and objectives of the organisation. Another important point in terms of structure is the concept of line and staff functions. Line functions are those involved in creating, developing, and delivering a programme. Staff functions are those that are of an advisory and consultative order. Line functions contribute directly to the attainment of the organisation's objectives, and staff functions contribute indirectly.

Staffing

A key aspect of managing an adult and extension enterprise is to find the right people for the right jobs. Much of one's success as a manager is related to appropriate human resource planning, regardless of whether it is the hiring of a secretary or an instructor for a particular work-shop. The staffing function consists of several elements:

1. *Human resource planning*—how many staff resources, with what backgrounds, and at what cost can be considered for objectives implementation?
2. *Recruitment*—how does one proceed to find the person with the appropriate mix of education, experience, human relations skills, communications skills, and motivation? An important component of the recruitment process is writing the job description. The description must be exact and specific but sufficiently general to solicit interest among potential candidates. The nature of the job, scope, authority, and responsibilities form the core of the job description. Indications of preferred educational background as well as salary range must also be included. In times of high unemployment, one can always expect several dozen applications for any one opportunity for employment. This leads to the next task of staff selection.

Staff selection

The process of staff selection involves evaluating candidates through application forms, curriculum vitae, and interviews and choosing the best candidate for the specific job responsibility. One can even have a list of criteria and a score sheet for each individual. Even then, successful hiring is often a very intuitive act and involves some degree of risk. As a means of giving some structure and design to the staffing process, the following guidelines are useful.

Staff orientation

This is the process of formally introducing the selected individual to the particular unit, to colleagues, and to the organisation. The selected person should be aware of the mission and objectives of the unit, the nature of responsibilities and level of authority, the degree of accountability, and the systems and procedures followed to accomplish the tasks associated with the job. A motivated individual will simply ask for such things as personnel manuals, administrative procedures handbooks, and aims and objectives statements. Such orientation tools should be available.

DIRECTING

At one time there was a management emphasis on "directing" in the directorial (autocratic) sense, but in recent times, the concept of directing has become more congruent with leading than with pushing. Thus today, directing is more related to leading and leadership styles. Leadership in this context means the process whereby a work environment is created in which people can do their best work and feel a proprietary interest in producing a quality product or service.

McGregor proposed that managers might assume that employees are motivated in one of two ways. His dichotomy was labelled theory X and theory Y. Theory X relates to traditional management whereby managers assume that they must control, coerce, and threaten in order to motivate employees. Theory Y, the opposite of theory X, suggests that employees want to do challenging work, that they are interested in accepting responsibility, and that they are basically creative and want to be involved in policy development and objective setting. Today, theories X and Y don't really sound very revolutionary; the problem is that management styles and employee motivation do not fit easily into two theoretical labels. As a result, additional theoretical labels are being developed.

COORDINATING

This important stage consists of interrelating the various parts of the work. It involves coordinating the various job roles and responsibilities of yourself and other staff, of your unit and other units within the same organisation, and of your unit with the broader community.

There are two forms of coordination:

(1) vertical reporting to your supervisor(s) and to your staff, and

(2) horizontal reporting to your colleagues and your management team.

Adult and extension educators are usually involved in very complex organisations such as governments, colleges and universities, and boards of education. Because of the size of the organisation, the increasing demands for public accountability, the many government regulations and policies, the increasing competition among providers of adult education opportunities, and the changes in technology, it is essential that the coordinating role be given top priority. How, then, can effective coordination be accomplished?

- Coordination needs professional, competent leadership, a democratic style that leads to trust, open communication, and ease of information flow.
- Coordination needs a constant definition and communication of mission and objectives that are understood by all managers.
- Coordination, to be effective, must have open, two-way channels of communication.
- Coordination involves a sharing atmosphere as well as commonly agreed on direction. Because effective coordination requires cooperation and communication, the meeting technique is still the most effective format for assuring the interrelationships among the various job responsibilities.

In recent years, formal systems of community coordination of adult and extension education activities have been developed. Such councils of continuing education provide more than just a network of workers but in fact lead to discussions of community needs, agency priorities, and an agreement as to who is going to look after what. Such voluntary coordination does not eliminate competition; it focuses on the multiway flow of information.

REPORTING

This function, closely related to the coordinating function, consists of keeping those to whom you are responsible informed as to what is going on. It is essential that competent managers keep the information flowing, especially in this age when there is so much information being transmitted in so many forms. The reporting function is more than preparing an annual report, quoting statistics, and informing your staff of current developments.

The reporting function is almost an evaluation function since it compares how you are doing with what you set out to do. It reviews your objectives and determines to what extent you are meeting your objectives. It consists of more than course numbers or annual statistics, but relates programme direction, policy changes, refinement in objectives, and changes in structures and priorities.

One of the key elements of the reporting function is the annual report. Such a report gives you the opportunity to summarise programmes, projects, and activities and to provide statistics as well. Such a report can be used as a public information document by having it distributed to other adult education agencies in the community, to your senior levels of management, to your own managers, to your colleagues, and to the press. In addition, it will prove to be a valuable document to satisfy the requests you receive asking about your programme activities.

BUDGETING

This management function includes fiscal planning, accounting and revenue, and expense controls. Budgeting requires specific planning, a thorough understanding of objectives and future programmes, a sixth sense of economic conditions and realities, and a hunch for predicting the unpredictable. In many cases, an organisation specifies the budget system being used. It could be based on

(1) historical data (what you had last year with variations for the coming year);

(2) 0-based data where the budget is created and justified on a line-item basis according to programmes and priorities;

(3) an MBO system—management by objectives whereby specific objectives are funded; and

(4) a PERT system—programme review and evaluation technique—where each programme is reviewed and assessed according to its contribution to specific goals. These are only a few of the budgeting systems in use.

However, the key elements of any budget system consist of (1) determining what line items are necessary in terms of objectives; (2) in line with policies, determining the financial amounts for each line; (3) determining overhead, surplus, and/or profit margins; (4) determining anticipated revenue from fees, grants, gifts, contracts, etc.; (5) drafting a budget with specific amounts and justifications; and (6) discussing and making adjustments to produce a working budget.

The budget then becomes a guide which, however, may always be in a state of change. The budget process is not in a vertical something that one does only once a year; it is a continual process of regular review and possible revision. One should always be checking to see how one is doing compared with how one anticipated doing.

Budget management, then, consists of three parts:

(1) budget determination—allocating revenue according to priorities and by line items;

(2) budget accountability—how well the anticipated budget matches reality; and

(3) using a +, 0—notation in answering the questions and by placing the notations in the boxes on the chart.

In this way, one can get a picture of the predominant types of management modes currently being used. While this may be useful in describing what is, it could be even more useful in describing what could be. It is also useful in providing some clues as to possible areas of role conflict—the scholarly research model would likely collide with the competent practitioner model. Structure is the basis for many modern business organisations because we live in a structured society, although the concept of structural rigidity and hierarchy is now being challenged by a more educated, creative, and intrinsically motivated workforce. The structural approach shows graphically that the organisation has a distinct physical shape or form provided by an internal form. A competent manager in this system is able to solve problems, and then enlist whatever support is needed to get it done. This approach is favoured by traditional, hierarchical, job-specific, uncreative organisations.

A more organic management method is based on paradigms. A paradigm refers to a method of approaching a problem or situation and the kinds of assumptions, values, and attitudes associated with thinking about the situation. It connotes a pattern or structure that is dynamic, changeable, and responsive to the environment. The most dramatic illustration of a paradigm shift was the shift from the Ptolemaic theory, which saw the earth as the centre of the universe, to the Copernican theory, which saw the sun as the centre of the universe.

A paradigm shift results in a total restructuring in the ways we think about a situation and the kinds of assumptions we make about former observations. Covey speaks of paradigm shifts: things, people, and structure can and do change—nothing is constant. He shows how almost every significant breakthrough is first a break with tradition, with old patterns, with old ways of thinking, and with old paradigms.

The use of models is common practice in management thinking. Models are useful because various aspects of the structures can be viewed from different positions that can then lead to new perspectives. The goal of modelling is to achieve an accurate yet relatively simple representation of a system, complex entity, or reality usually on a smaller scale. It implies both structure and change. An important aspect of model building is collecting and preparing data. Information gained from the data is the foundation of the model. The model should reflect the major aspects of the problem as simply as possible. Often, this requires tradeoffs because simplicity and accuracy rarely go hand in hand. The constants, if any, within the model should be known with a high degree of precision. One mistake that inexperienced model builders often make is failing to take a broad perspective of the problem. They do not take into account other dimensions of reality that a solution may have an impact on. To accomplish this broad perspective, the extension manager should adopt a systems approach to model building and should focus not only on the immediate problem, but also on interrelationships that exist within and outside the organisation and how these relationships will be affected.

MODERN MANAGEMENT APPROACHES

Systems Theory

Modern management is characterised by two approaches, the systems and the contingency approach. The systems approach views the organisation as a total system comprised of interacting subsystems, all of which are in complex interaction with the relevant external environment. Organisations are pictured as "input-transformation-output systems" that compete for resources. The survival and prosperity of an organisation depend on effective adaptation to the environment, which means identifying a good strategy for marketing its outputs (products and services), obtaining necessary resources, and dealing with external threats.

Survival and prosperity also depend on the efficiency of the transformation process used by the organisation to produce its goods and services, on worker motivation, and on cooperation. Efficiency of the transformation process is increased by finding more rational ways to organise and perform the work and by deciding how to make the best use of available technology, resources, and personnel. Top management has primary responsibility for designing an appropriate organisational structure, determining authority relationships, and coordinating operations across specialised subunits of the organisation. A system can survive only when it delivers an output that can be exchanged for new inputs as well as for maintaining the system. Similarly, an extension service is expected to produce some beneficial output.

Contingency Theory

Theories that explain management effectiveness in terms of situational moderator variables are called contingency theories. The contingency or situational approach recognises that neither the democratic nor the autocratic extreme is effective in all extension management situations.

REFERENCES

Chambers, R. 1993.. *Challenging the profession - Frontiers for rural development.* London: Intermediate Technology Publications, p. 67.

Cherrington, D. J. 1989.. *Organizational behavior: The management of individual and organizational performance.* Boston: Allyn and Bacon.

Claar, J. B., & Bentz, R. P. 1984.. Organizational design and extension administration. In B. E. Swanson Ed.., *Agricultural extension: A reference manual.* Rome: FAO.

Clark, R.C. 1973.. *Basic concepts and theories of administration and supervision* revised edition.. Madison:

Elkins, A. 1980.. *Management structures, functions, and practices.* Massachusetts: Addison-Wesley, p. 539.

Evans, J. P., & Dahl, D. T. 1984.. Organizing for extension communication. In B. E. Swanson Ed.., *Agricultural extension: A reference manual.* Rome: FAO.

Flemming, P. 1989.. *Strategic planning: Fact sheet.* Ontario: Ministry of Agriculture and Food, Rural Organizations and Services Branch.

Ivanevich, J. M., Donnelly, J. H., Jr., & Gibson, J. L. 1980.. *Managing for performance.* Georgetown, Ontario: Irwin Dorsey.

McNeil, A., & Clemmer, J. 1988.. *The V.I.P. strategy-leadership skills for exceptional performance.* Toronto: Key Porter Books Ltd.

15

Human Resource Planning for Extension Services

One of the most significant developments in the field of organisation in recent times is the increasing importance given to human resources. More and more attention is being paid to motivational aspects of human personality, particularly the need for self-esteem, group belonging, and self-actualisation. This new awakening of humanism and humanisation all over the world has in fact enlarged the scope of applying principles of human resource management in organisations. The development of people, their competencies, and the process development of the total organisation are the main concerns of human resource management.

Extension organisations in developing countries face the major problems of professional incompetence and lack of motivation among their employees. Further, many of the agricultural extension departments of these countries do not have a well-defined system of human resource management. Proper planning and management of human resources within extension organisations is essential to increase the capabilities, motivation, and overall effectiveness of extension personnel.

Keeping this in view, the various dimensions of human resource management as applicable to extension organisations: human resource planning for extension, job analysis, recruitment and training of extension personnel, performance appraisal, supervision, management of rewards and incentives, improvement of the quality of work life, and organisational development for extension.

Human resource planning forecasts the future personnel needs of extension organisations. With the rapid changes in technology, needs of farmers, market situation, and competitive environment, planning for human resources has become an important, challenging task for extension. Human resource planning involves plans for future needs of personnel, their required skills, recruitment of employees, and development of personnel.

Human resource forecasting and human resource audit are the two most important components of this type of planning. Human resource forecasting refers to predicting an organisation's future demand for number, type, and quality of various categories of employees. The assessment of future needs has to be based on analysis of present and future policies and growth trends.

The techniques of forecasting include the formal expert survey, Delphi technique, statistical analysis, budget and planning analysis, and computer models. The human resource audit gives an account of the skills, abilities, and performance of all the employees of an organisation.

JOB ANALYSIS

Job analysis traditionally was done for purposes connected with recruitment, pay, administration, and supervision. But the increasing complexity of work has made job analysis an important instrument for developing people in organisations. Job analysis requires a systematic collection, evaluation, and organisation of information about the job. This information is collected through interviews, mailed questionnaires, observation, study of records, and similar methods. The collected information becomes a basis for preparing job descriptions and specifications. The job description, or job profile, is a written statement which includes detailed specifications of duties to be performed, responsibilities, and working conditions and indicates what is expected of a job holder. A job specification is a profile of the human characteristics needed for the job, such as education, training, skills, experience, and physical and mental abilities. Extension organisations in developing countries do not have clearly defined job descriptions or job specifications for extension personnel.

The training and visit system of extension considerably improved the preparation of job charts, work plans, and time-bound work for different categories of extension personnel. However, the actual utility of job descriptions in extension organisations is complicated by factors such as work overload, seasonality of extension, the range of cropping systems, and distribution of extension service over a large area.

Studies analysing the role of extension agents reveal that they face work-related problems such as role ambiguity and lack of job authority, expertise, and accountability. This shows that job analysis is needed to improve the performance and effectiveness of extension employees. Job analysis can more effectively contribute towards the development of extension personnel by adopting the following procedures which involve identifying key performance areas (KPAs) and critical attributes.

Key Performance Areas

A job description consists of many details, but does not specify key areas which need attention. Further, it gives the details of what is expected from the current jobholder. On the other hand, key performance areas are specific and show the critical functions relevant at present and for the future to achieve the objectives.

The identification of key performance areas helps in role clarity as well as in delegation of functions. This in turn aids in performance appraisal and training. Generally, four or five key areas for a job are identified. The core extension personnel of developing countries consists of village extension workers, subject-matter specialists, and supervisory staff or extension officers. Examples of key performance areas of core extension personnel are given below.

Village Extension Workers. People in this category:

(1) make regular and systematic visits to villages and farms to develop rapport with the clientele and to understand their problems;

(2) undertake educational activities in the form of meetings, campaigns, demonstrations, field days, training sessions, and exhibitions; and

(3) provide advisory services to the farmers and solve their production problems.

Subject-Matter Specialists. Their role is to:

(1) keep abreast of current recommendations and findings related to farm production by maintaining continuous contact with agricultural research stations;

(2) provide feedback to the research system about farmers' problems which need solutions; and

(3) train and backstop village extension workers on the latest farm technology and help them in solving field problems.

Supervisory Staff or Extension Officers. People holding these positions:

(1) plan, organise, coordinate, and implement extension programmes and activities;

(2) supervise and monitor the work of field staff, providing guidance, motivation, and evaluation of performance; and

(3) coordinate the programme with inter-and intradepartmental agencies.

Critical attributes for extension personnel

The key performance areas indicate the important roles and contributions of different categories of extension personnel. Once the roles are delineated, they can be analysed to indicate the attributes which can discriminate an effective from an ineffective role occupant. These critical attributes consist of qualities such as educational qualifications, skills, experience, physical characteristics, mental abilities, values, and attitudes needed for extension.

The critical attributes needed for field-level and supervisory extension staff are necessary formal training in agriculture, practical skills and experience in farming, and knowledge of modern farm practices. Abilities in group dynamics, human relations, and communication are also important. Basic skills related to management and leadership are needed by extension supervisors. Values and attitudes such as faith in rural people, commitment to agricultural development, and concern for the whole community are important for all extension personnel.

The importance of assessing personal and professional attributes for selecting productive extension personnel has been reported by several researchers. Assessment is essential because an unsatisfactory educational level of extension staff is one of the most serious problems of extension in countries like Bangladesh, Botswana, Kenya, Malaysia, Sudan, and Zambia. A worldwide analysis of the status of agricultural extension reveals the low level of formal education and training of field extension agents in developing countries.

RECRUITMENT AND TRAINING OF EXTENSION PERSONNEL

Recruitment is important in selecting the right kind of extension personnel. Since the job of extension personnel calls for technical skills as well as commitment and willingness to educate rural people, an appropriate selection system is essential to ensure the right selection. The success of extension depends heavily upon selection of qualified and motivated personnel. Extension organisations in developing countries use two major sources of recruitment: from outside and from within. Entry-level positions such as village extension workers and agricultural extension officers are filled by outside recruitment, using the services of government placement agencies.

Other channels of recruitment are advertisements, private placement agencies, professional search firms, and educational institutions. In some countries, farmers are recruited to help extension agents. In Israel, volunteers with practical experience in farming, usually a couple, were recruited as extension workers to help the immigrants. These agents were found to be enthusiastic; they lived with the farmers, set a personal example, and were effective instruments for making desired changes. Most of the extension departments in developing countries have the policy of promoting or recruiting within for middle-level and top-level positions.

In India, positions like deputy director, joint director, and additional director of extension are filled through promotion. The advantages of this policy are that it promotes loyalty and provides opportunities for existing extension staff to get high-level positions.

However, its greatest disadvantage is that it prevents the lateral entry of talented extension personnel and promotes complacency because seniority ensures promotion.

Techniques for Selecting Extension Staff

The selection of extension staff starts with making the job opportunities known to all potential applicants through advertisement. The help of extension workers' training centres, agricultural colleges, rural institutions, and local government agencies may be sought to give wide publicity, as well as to inform candidates living in rural areas. This is followed by screening applicants to short-list suitable candidates and by evaluating potential candidates through various tests.

A typical selection process consists of the following steps: completed job application, initial screening, testing, indepth selection interview, physical examination, and job offer. In general, extension organisations in developing countries use a simple knowledge test and a brief interview to select extension personnel. By using the above method, it is impossible to discriminate an effective candidate from an ineffective candidate, because selecting extension personnel demands thorough, indepth testing of cognitive and noncognitive abilities. Testing cognitive ability includes a knowledge test, a skill or ability test, and an aptitude test.

A noncognitive test is a measure of behavioural dimensions which are important for field-level extension personnel, including concern for and commitment to rural people, empathy, problem-solving orientation, high motivation to influence and educate farmers, ability to work under unsupervised and difficult village conditions, patience and persistence, and team spirit. A good example of selecting village-level extension workers on the basis of behavioural characteristics is provided by the extension project of Allahabad Agricultural Institute.

In response to an advertisement for 27 posts of village guides, 700 to 800 candidates had applied. The final selection procedure consisted of five days of testing skills and attitudes in actual village situations. The test included testing attitudes towards menial tasks like cleaning a cattle shed or digging a compost pit. The candidates' responses to emergency situations were also tested by dropping them into isolated villages.

The assessment centre approach, originally used during World War II, can be used to select extension staff. In this approach, an organisation develops its internal resources for assessing new staff. The candidates to be recruited go through a number of simulation exercises, and an expert assesses their behaviour. The techniques used are a psychological test, role play, in-basket exercise, group discussion, projective test, knowledge test, and interviews.

TRAINING AND DEVELOPMENT

The training of extension personnel contributes directly to the development of human resources within extension organisations. "Training programmes are directed towards maintaining and improving current job performance, while development programmes seek to develop skills for future jobs". Training has to start with the identification of training needs through job analysis, performance appraisal, and organisational analysis.

Once the training needs of extension personnel have been identified, the next step is to organise training programmes. Methods such as games, role playing, simulation exercises, and case study can be used in extension organisations to create learning situations based on experience. Training based on actual field experience should be emphasised. Emerging new farm technologies such as integrated pest management and improved practices in horticulture call for actual field experience. Extension agents need training not only in the technological aspects but also in human relations, problem solving, sensitivity towards disadvantaged groups, and the basic concepts of management.

Management Development Programmes

Management development programmes are meant to improve the managerial skills of senior-level extension officers and to prepare them for future roles. There is a great need for management development programmes in extension organisations because they face complex situations due to changing agricultural scenarios. Further, extension managers have to be exposed to modern management techniques and methods.

Management development programmes have to be suited to the needs of top-level extension managers and should be based on needs analysis. Methods such as coaching, job rotation, training sessions, classroom instruction, and educational institute-sponsored development programmes are used to train managers. In India, a separate institute called Manage has been established to train senior extension managers in managerial skills and human relations.

PERFORMANCE APPRAISAL

How extension personnel are recruited and trained and become part of a work group. These are all vital activities. However, the ultimate measure of effective human resources within an extension organisation is the performance of extension personnel. Thus performance appraisal is important for effective human resource management.

Performance appraisal is a process of evaluating employee performance in order to guide and develop the employee's potential. In many extension organisations which are government departments, the performance appraisal is nothing more than a confidential judgement of work done and a character report used to facilitate disciplinary action or promotion. The employees do not get feedback about their performance.

Extension organisations need to have an open appraisal system to provide feedback and opportunities for open discussion with employees on their performance, because they have immense potential to grow and develop. This system can create a healthy working climate and employee motivation. The performance appraisal which aims at facilitating employee development has the following major purposes:

(1) to provide feedback and guidance,

(2) to set performance goals,

(3) to identify training needs, and

(4) to provide inputs for management of pay administration, rewards, and promotion.

The steps involved in effective performance appraisal are

(1) identification of key performance areas and setting yearly objectives under each KPA,

(2) identification of critical attributes for effective performance,

(3) periodic review of performance,

(4) discussion of performance with employees, and

(5) identification of training and developmental needs.

Potential Appraisal

The potential appraisal is a future-oriented appraisal by which the potential of an employee to occupy higher positions and to assume higher responsibilities is evaluated. The potential appraisal can help the extension staff to know their strengths and weaknesses and can motivate them to further develop their skills. Thus the potential appraisal helps in planning overall career development of employees. Some of the techniques used for the appraisal are self-appraisals, peer rating, the management by objectives (MBO) approach, psychological test and simulated work exercises, case analyses, and leadership exercises.

Performance Review and Counselling

An important purpose of the performance appraisal is to counsel and guide employees towards greater job effectiveness. Thus a system of performance counselling is needed in extension organisations. Performance counselling is provided by the manager to the subordinates to help them in the analysis of job performance, identification of training needs, and finding solutions to the problems which hinder job effectiveness. Counselling is an art of communication involving two people - manager and employee. Counselling differs from training in that the former involves a dyadic relationship and establishes more mutuality and confidentiality. The success of performance counselling depends upon the employee's interest, a climate of openness and mutuality, and the counselling process. Extension managers can use directive, nondirective, and cooperative counselling.

SUPERVISION

Two major functions of supervision are task orientation and concern for employees. Therefore, direction and organisation of activities, motivation of employees, and management of work groups are the important functions of extension supervisors.

Direction and Organisation

Extension supervisors have to plan the work and maintain a high standard of performance. The whole process of job analysis, identification of key performance areas, and performance appraisal will help in planning and organising extension work. The training and visit system of extension has introduced mechanisms for defining goals, planning, and scheduling work at the field level with provisions for monitoring and evaluation. Some of the management techniques used by extension organisations in overall planning and management of programmes are the programme evaluation and review technique (PERT/critical path method (CPM), management by objectives (MBO), programme and performance budgeting system (PPBS), and time management techniques. These techniques have been practised by extension organisations in Asian and African countries with varying success. Personal computers offer good scope for extension managers to increase certain managerial skills.

A study of supervisory practices to improve field performance of agricultural extension in Kenya, Malawi, and the Philippines revealed the following effective supervisory practices:

(1) use of collaborative, realistic, and result-oriented target setting and a daily activity plan;

(2) a needs-based participatory evaluation system;

(3) involvement of farmers in decision making and a reachable service target under local constraints; and

(4) effective communication and use of simple proforma and report procedures.

Studies of agricultural extension in Asia and Africa show that extension supervisors must be considerate as well as task oriented, involving subordinates in decision making and treating the employees with more interpersonal competence. Up to 86 per cent of field agents in Southeast

Asia reported "friendship" as the most effective way for supervisors to ensure extension workers' reliable performance. Thus if extension managers are to be effective, they have to give supportive evaluation by way of enhancing employee motivation and improving the functioning of work groups.

MOTIVATING THE EXTENSION PERSONNEL

The work motivation and morale of extension staff, as reported earlier, are very poor in many countries. The reasons are many. The bureaucratic structure of extension administration, lack of rewards and incentives, poor facilities, poor promotional avenues, and the low esteem given to extension are the major causes of poor motivation and morale. Extension supervisors should have the ability to motivate and lead the field extension workers so that the field agents perform more than routine jobs, and supervisors should be involved in attaining excellence in extension work. This calls for extension managers having an understanding of various theories of motivation as applicable to frontline extension agents. Therefore, a knowledge of major theories of motivation such as Maslow's hierarchy of needs theory, Herzberg's two factory theory, McClelland's need theory, theory X and theory Y, and expectancy theory of motivation is essential. Special training for developing motivation among field-workers has to be undertaken by supervisors.

WORK-GROUP MANAGEMENT

Every organisation has formal and informal groups. Formal groups are established by the management, while informal groups are spontaneous and developed to satisfy mutual interest of the members. Because work groups have a considerable influence on the work situation, supervisors should be sensitive to the needs of the group and develop skills to guide and achieve the group's goal, which will benefit the organisation and the members. Effective extension supervision can use work groups in problem solving because they can provide many creative solutions.

One way to improve supervisory effectiveness in extension work is to develop a leadership style which represents the extension workers' group interest at the higher level of organisation. This will increase the confidence and morale of the work group. An understanding of group dynamics and their implications for increasing work-group performance is essential for extension supervisors. For example, in the "Hawthorne Effect," increased performance due to special treatment of the group can be effectively used in extension organisations. Studies have pointed out that well-developed group dynamics result in increased extension performance.

REWARDS AND INCENTIVES

An important aspect of human resource management which needs special attention in extension organisations is the development of a reward system which will attract, retain, and motivate extension personnel, as well as provide training and promotional opportunities. Extension organisations in Asian and African countries have a poor reward system. The extension agents are not only poorly paid but are paid late and after reminders or visits to head-quarters.

Most of the extension services are run by government agencies and operate under rules and regulations of public administration. These rules do not have provisions for rewarding superior performance or for a wage system based on merit. Promotion criteria are based on seniority and length of service. Thus the bureaucratic structure of extension services is a basic hindrance to designing a better reward system. Among many of the government departments, the agricultural department and extension service have a low public esteem and poor pay structure. The rewards and incentive system can be improved in several ways.

Rewarding Superior Performance

Extension organisations have to develop a reward system which encourages superior performance so that pay and wage administration will be an effective tool to promote performance, motivation, and satisfaction. A clear job description, performance standards, and performance appraisal will help in evaluating extension work and rewarding people for meritorious service. Ways and means have to be found within the existing framework of public administration for basing pay on performance. For example, extension workers on the basis of their performance can be sent for higher education. Nonmonetary rewards such as recognising the good ideas of field workers or awarding honourable titles will also help in improving performance. Extension personnel may also be encouraged to form professional societies to develop and communicate high standards, as well as to recognise superior performance. A professional monthly journal or newsletter can help extension agents to communicate innovative ideas and reinforce superior performance.

Improved Working Conditions at the Field Level

The reward system must also be internally equitable. The relative importance of field-level extension functionaries has to be realised in terms of pay compensation and other amenities. Lower level extension workers often have to work under unpleasant and isolated conditions. A carefully planned system of field allowance will compensate this. The living conditions of field extension workers must be improved by providing adequate facilities for housing, transport, and medical and educational allowances for children.

Career Planning and Development for Extension Personnel

A career refers to all of the jobs that people hold during their working lives. Career planning is the process by which employees plan their career goals and paths. Career development refers to all of the technical and managerial skills employees acquire to achieve their career plans. Career advancement, which gives a picture of future opportunities in terms of promotion, is a motivating factor for performance and development of skills.

Unfortunately, no career structure exists for extension personnel in many organisations. In developing countries like India, there are many cases where one joins as a village extension worker and retires in the same position after serving thirty to thirty-five years. As part of improving the rewards and incentives system, extension organisations have to develop suitable career paths and advancement for different categories of extension personnel on a systematic basis.

As part of career development, extension personnel should be provided with opportunities to develop their technical and managerial skills to enable them to occupy higher positions.

Extension personnel should have a salary structure as well as promotion opportunities comparable to other professions like health or engineering. In Kenya, the pay and career opportunities of extension workers are comparable to other government employees. Recommendations have already been made to equate the status of agricultural extension with that of agricultural research by offering an equal salary structure, professional advancement, and incentives and rewards.

IMPROVEMENT OF THE QUALITY OF WORK LIFE

The earlier approach to human resource development emphasized individual development through training and proper supervision. However, with the increasing complexity of organisations and society, it was soon realised that training individuals plays only a limited role in the development of organisations. The need for improving the quality of work life through making the job more satisfying and productive has been greatly felt. Factors such as the nature of the job or the role and involvement of employees in work decisions are important for improving the quality of work life.

The methods used to do so are job enrichment, job design, and role interventions. An understanding of these methods and their application in extension organisations are essential for extension managers to improve the performance of extension agents. Studies have shown that the work environment of extension organisations is poor and needs improvement.

JOB ENRICHMENT AND JOB DESIGN

Job enrichment refers to detailed analysis of the work to know the factors which make it a satisfying experience. Job enrichment uses the job as the medium of developing employees and changing organisational practices. Some of the factors which increase job satisfaction are a sense of achievement in the job, recognition for the job, the nature of the work itself, and opportunities to learn new things and grow.

The principles of job enrichment, according to Herzberg, are removing controls while retaining accountability, introducing new tasks, giving a complete unit of work, granting job freedom, and helping employees to become expert in their tasks. These principles can be practised by extension managers to increase the quality of work and job satisfaction among extension personnel. Job enrichment programmes were successful in improving the quality of work and job satisfaction. However, it was found that job enrichment had a limited view of the job, and the need for greater emphasis on human values was realised. This led to the concept of job design, which refers to structuring a job to satisfy the technical, organisational, social, and human requirements of the person performing the work.

Based on the humanisation of work, job design aims at increasing the quality of work life through treating the employees as human beings and emphasizing their development and involvement in work decisions. It emphasizes the use of extrinsic and intrinsic job factors, employee participation in management, autonomy, adaptability, and variety. The concept of job design can be used by extension managers to increase participation of extension personnel in the planning and management of extension programmes, which will improve the quality of their work life.

ROLE INTERVENTIONS

The study of roles, which are the positions employees hold in an organisation, as defined by the expectations of significant persons and the individuals occupying the positions, is a comparatively neglected aspect of organisations.

Roles are an important dimension in increasing organisational effectiveness. Through their roles, people are linked with the organisation. This linkage increases organisational effectiveness by integrating the individuals with the organisation. Such integration increases mental well-being and personal effectiveness. The purpose of role-based intervention is to increase the mutuality of roles in organisations.

Role-based interventions are done through learning situations such as process laboratory, group discussion, and use of questionnaires and schedules. Role-based interventions in extension organisations will result in increased work commitment, motivation, creativity, and team spirit.

ORGANISATIONAL DEVELOPMENT

An efficient extension organisation needs to develop the capability of responding to changes in relation to its environment. Extension organisations have to cope with changes within and outside the organisation, such as changes in farm technology, communication methods, needs of farmers, rural situations, export and import of farm produce, and market economy. Organisational development allows for planned changes in the organisation's tasks, techniques, structure, and people. Attitudes, values, and practices of the organisation are changed so that it can cope with changing situations. The employees also gain greater skills to deal with new problems.

Also focussing on team building and conflict management, organisational development is a planned effort and is done with the help of an external expert in the behavioural sciences. The process consists of diagnosis of the problem, data collection, feedback of data to the organisation, introduction of specific interventions, evaluation, and follow-up. Techniques such as sensitivity training, transactional analysis, and team-building exercises are used to develop interpersonal relationships. Organisational development is an effective approach that can be used by extension organisations to bring about planned changes and to increase the interpersonal relationships among the employees.

REFERENCES

Adams, M. E. 1982.. *Agricultural extension in developing countries.* Burnt Mill, Harlow, Essex: Longman

Bathgate, F. Ed... 1956.. *Experiment in extension-The Goon Sathi.* London: Oxford University Press.

Baxter, M. 1990.. Agricultural extension investment. In D. C. Misra Ed.., *Training and visit system of agricultural extension in India in action.* New Delhi: Ministry of Agriculture, Government of India.

Bhasin, K. 1976.. *Participatory training for development.* FFHC/AD. Bangkok: FAO.

Blanckenburg, P. V. 1984.. *Agricultural extension systems in some African and Asian countries.* FAO Economic and Social Development Paper 46. Rome: FAO.

Blum, A. 1987.. The Israeli experience in agricultural extension and its application to developing countries. In W. M. Rivera & S. G. Schram Eds.., *Agricultural extension worldwide - Issues, practices and emerging priorities.* London: Croom Helm.

Chattopadhyay, S., & Pareek, U. 1982.. *Managing organizational change.* New Delhi: Oxford & IBH Publishing Company.

Davis. L. E. & Taylor, C. Eds. 1979., *Design of jobs.* Santa Monica, CA: Good Year.

FAO 1985.. *Report of the expert consultation on linkages of agricultural extension with research and agricultural education.* Bangkok: RAPA/FAO.

French, W. L. 1982.. *The personnel management process* 5th ed... Boston: Houghton Miffin.

Goodell, G. E. 1983.. Improving administrators' feedback concerning extension, training and research relevance at the local level: New approaches and findings from South-east Asia. *Agricultural Administration, 13,* 39-55.

Gupta, T. S. 1963.. *Characteristics of effective village level workers.* Unpublished master's thesis, New Delhi, Division of Agricultural Extension, Indian Agricultural Research Institute.

Hayward, J. A. 1990.. Agricultural extension: The World Bank's experiences and approaches. In B. E. Swanson Ed.., *Report of the global consultation on agricultural extension* p. 115-134.. Rome: FAO.

Herzberg, F. 1966.. *Work and nature of work.* Cleveland: World Publishing.

16

Monitoring Extension Systems

The Global Consultation on Agricultural Extension observed that monitoring and evaluation are important yet frequently neglected functions in most organisations. In the worldwide survey of national extension systems, it was found that only about one half of all national extension systems have some type of monitoring and evaluation capacity. The consultation noted that in many cases the M & E units are weak and are limited to ad hoc studies. Frequently, these M & E units are abandoned when project funding terminates.

In many organisations, monitoring and evaluation have a negative image because these units may concentrate on problems, exposing weaknesses and failures. Instead, monitoring and evaluation should be used in a positive manner to improve extension's performance and increase its efficiency. Therefore, attitudes about and uses of M & E must be changed if this capacity is to be used to advantage in strengthening extension's performance and impact. The consultation recommended that national extension systems should be strongly encouraged to establish and use monitoring procedures and evaluation studies both to improve extension performance and to communicate the results of extension programmes to policy makers and clientele being served.

Extension systems need to carefully consider monitoring, management information needs, sources of information, and a management information system. The word "monitor" is derived from the Latin word meaning to warn, and "evaluate" stems from the word value.

Monitoring is an integral and important part of a management information system. Managers require information to keep track of an extension programme and to guide its course of action.

Management information needs include six kinds of information:

(1) diagnostic information (why a situation is as it is),

(2) implementation information (physical and financial or input information),

(3) utilisation information,

(4) impact information,

(5) situation information, and

(6) information for review.

Sources of information cover a spectrum of sources which supply information to management, ranging from informal to formal sources. Management should be ever alert to receive feedback on extension programmes - whatever the source. Monitoring, it is useful to realise, is only one of management's sources of information.

A management information system is a scheme by which the "right" information is obtained in the right amount at the right time and is made available to the right person or persons.

An information system is usually created in modem organisations to cater to the information needs of management. A monitoring system is a subsystem of a management information system and has several distinguishable components.

CONCEPTUAL FRAMEWORK FOR MONITORING

A conceptual framework for monitoring consists of four principal components: an extension organisation, a monitoring and evaluation (M & E) unit, an information needs matrix, and a monitoring and evaluation cycle. Top management receives information from the monitoring unit and from other formal and informal sources. This influences programme implementation, leads to better programme planning, and ensures sustainability of extension programmes. Ultimately this leads to institutional development, which has been defined as "the process of improving the ability of institutions to make effective use of available human and financial resources".

APPROACHES TO EXTENSION MONITORING

Several approaches to extension monitoring are available. They arose, notably during the 1980s, as a response to the need to improve extension performance. Basically similar, these approaches, however, differ significantly in their emphasis. All of them advocate simplicity and timeliness, essential requirements of good monitoring.

Traditional (administrative) Approach

Based on routine administrative reporting, this approach is concerned with physical and financial achievements in a programme. Its primary weaknesses include multiplicity of reports by programme personnel and absence or neglect of beneficiary contact. It is being increasingly replaced by other approaches.

Zones-of-concentration Approach

The Cernea and Tepping approach, introduced in 1977, concentrates on three zones:

(1) visits, as the final outcome of extension efforts;

(2) recommendations, as the content of the visit and means towards the end yields; and

(3) yields, as the eventual consequence of the development effort.

Methodological Approach

The Slade and Feder approach, introduced in 1981, builds upon the zones-of-concentration approach and suggests a monitoring survey early in each cropping season, a monitoring-cum-evaluation survey in each cropping season at the time of harvest, specific indicators, and reporting. Working manuals are a characteristic feature of this approach.

Expanded Monitoring Approach

Casley and Kumar suggested in 1987 an expansion of the monitoring function to cover not only physical and financial information, but also beneficiary contact information and project diagnostic studies. Under this approach, there is greater emphasis on monitoring and less on evaluation. Project diagnostic studies are a novel feature of this approach.

Adoption Rates Approach

Murphy and Marchant suggested in 1988 an approach which concentrates on adoption rates as key indicators. This approach moves away from trying to monitor agricultural results and concentrates on directly monitoring the provision of and response to project services. Under this approach, extension service is treated as the "medium" and the recommendations as the "message."

Marketing Approach

Lee has suggested an approach which is based on market segmentation, a standard technique in marketing. Under this approach, the need and likely demand for new technology are first assessed, and then target market segments are predicted. For example, 10 per cent of the targeted farmers are likely to readily adopt a new technology; a further 15 per cent are likely to adopt the technology when returns for wheat reach x dollars per ton; 40 per cent of the targeted farmers are likely to adopt the technology when they perceive that their peers have adopted it; and 35 per cent of the target market is not likely to adopt the technology at all in the next five years. Among these approaches, Lee's approach has much to commend it, because it is based on careful assessment of the need for new technology, an exercise seldom under-taken by extension organisations.

OPERATIONALISING THE DEFINITION OF MONITORING

Monitoring is a specialised, dynamic, semiautonomous, and institutionalised management resource. Preferably it is computerised. Monitoring helps to ensure the implementation of extension programmes in accordance with their design and takes into account the interests of various stakeholders.

The definition of monitoring can be operationalised by establishing principles to follow; setting up a data collection system; establishing relationships among the monitoring unit, management, the extension staff, and extension's clients (farmers); and making appropriate use of nongovernmental organisations (NGOs).

PRINCIPLES OF MONITORING

In light of past experience in monitoring and its designated role, it is possible to lay down the following ten principles of monitoring:

1. *Monitoring must be simple.* A complex or complicated monitoring system is self-defeating. The basic task of monitoring is to simplify the field-level complexity, sifting the more important concerns from the less important.
2. *Monitoring must be timely.* Timeliness is of the essence in monitoring. Management requires input from the monitoring system so that timely action may be taken. Also, timeliness is closely related to the credibility of monitoring itself.
3. *Monitoring must be relevant.* It must concern itself only with parameters which are relevant to programme objectives. This also ensures that monitoring does not generate information that is not used or is not usable by management.
4. *Information provided through monitoring should be dependable.* Management will rely on monitoring findings only if the information is believed to be reasonably accurate.
5. *Monitoring efforts should be participatory.* Effort should be made to ensure participation by all concerned with extension, be they field-level personnel, subject-matter specialists, or extension's clients (the farmers).
6. *Monitoring must be flexible.* It is iterative in nature. It also gets routinised with the passage of time. These two features should not, however, lead to rigidity.
7. *Monitoring should be action oriented.* Monitoring often leads to action. Consequently, it should follow pragmatic approaches, keeping the requirements of extension's

clients uppermost in view. Generating information for which there is no intended use should be assiduously avoided.

8. *Monitoring must be cost-effective.* Monitoring efforts cost money and time. It is therefore essential to make it cost-effective. While principles of simplicity, time-lines, relevance, and accuracy will lead to cost-effectiveness, computerisation also can help to make monitoring more cost-effective by reducing staff hours in data processing.
9. *Monitoring efforts should be top management oriented.* Monitoring units should keep in mind the requirements of top management when designing and operating a monitoring system. Yet at the same time, monitoring must take into account the fact that those who provide information to the system also must benefit or the quality of the information provided will decline.
10. *Monitoring units represent specialised undertakings.* Monitoring is not merely concerned with the collection and analysis of data, but with diagnosing problems and suggesting alternative practical solutions.

STAKEHOLDERS OF MONITORING

There are many stakeholders in monitoring efforts. The funding agency, usually the ministry of finance or planning, wants to know that money has been spent properly. The ministry of agriculture wants to be able to say it has been efficient and effective. The target beneficiaries, that is, small and marginal farmers, want help from the organisation and want to improve the quality of their lives. And nongovernmental organisations (NGOs) need information generated by monitoring efforts and also want to contribute information. Thus there are four main stakeholders in monitoring - funding agencies, the implementing agency, beneficiaries, and NGOs - and their interests need to be taken into consideration when designing a system of monitoring.

FREQUENCY OF MONITORING

Monitoring is an ongoing, continual exercise. Data collection involving crop production should be undertaken twice in a cropping season: at the time of sowing to obtain benchmark information and again at the time of harvest. The period of recall for collection of data should not be more than a month.

MONITORING UNIT

The monitoring unit should be staffed by technical personnel having specialised skills. The staff often consists of extension specialists, economists, sociologist or anthropologists, statisticians, computer programmers, and supporting staff. The head of the monitoring unit may come from any of these disciplines. The staff should be no larger than necessary to accomplish the work of the unit. The leader of the monitoring unit should report to one of the top managers in the organisation.

MONITORING INDICATORS

Indicators, as the term suggests, are variables that help to measure changes in a given situation. They are tools for monitoring and evaluating the effects of an activity. Indeed, indicators are the principal means by which a monitoring unit keeps track of extension's capability, effectiveness, and efficiency. Any monitoring system will therefore incorporate the use of appropriate indicators in these four aspects.

There are two approaches to indicator development: the inductive and the deductive. In the inductive approach, a system of social, economic, and demographic statistics is created and a wide range of indicators is developed on the basis of the statistics available. This is the approach followed in the case of United Nations Social Indicators. In the deductive approach, the areas of interest are first identified, and then requisite indicators are developed. Socioeconomic indicators for the World Conference on Agrarian Reform and Social Development (WCARD) follow this approach.

In extension monitoring, both the inductive and the deductive approaches are followed; that is, indicators are developed on the basis of available statistics, and some are developed as a result of specially collected data.

There are different types of indicators, for example, development indicators, socioeconomic indicators, agricultural development indicators, and extension indicators. They range from general to specific concerns.

Indicators can be categorised into direct and indirect or proxy indicators; single and unitary or composite indicators; quantitative and qualitative indicators; primary, core, and supplementary indicators; input and output indicators; and monitoring and evaluation indicators. The criteria for selecting indicators depend upon the purpose, resources, and time available. The following criteria are usually suggested:

- *Simplicity:* The indicator should be simple enough to be understood by nonspecialists.
- *Unambiguous definition:* It should be clearly defined.
- *Ready determination:* The data can be obtained without undue difficulty. This is also referred to as "timely" and "feasible".
- *Accurate measurement:* The indicator should be measured accurately, which is often difficult when dealing with farming communities.
- *Validity:* The indicator should actually measure what it is supposed to measure.

- *Relevance:* It should be geared to the specific needs of decision makers and be relevant to project objectives.
- *Specifity:* It should reflect changes only in the situation concerned and should measure specific conditions that the project aims to change.
- *Consistency:* The value of indicators should stay constant so long as they are collected in identical conditions, no matter who does the collecting. Indicators should be objective and verifiable.
- *Sensitivity:* Indicators should be sensitive to changes in the situation being observed. They should be sensitive enough to reflect changes in the phenomenon.
- *Prioritisation:* Indicators should be prioritised and a minimum feasible list prepared.

BASIC CONCEPTS OF MONITORING

These four concepts are basic to monitoring and evaluation. They correspond, respectively, to operational investment (e.g., investment in agricultural extension per farm family), operational efficiency (e.g., the number of visits, meetings, demonstrations, and trials, per extension worker), technical efficiency (e.g., the number of adopters, hectorage, output, and value added), and extension-induced changes (e.g., production, productivity, income, and income distribution).

Capability, effectiveness, and efficiency fall in the monitoring domain. Impact falls in the evaluation domain.

Capability is the command that extension has over physical, financial, and human resources, enabling it to serve its clients. It is reflected by extension's outreach, intensity, technical competence, and physical and financial resources. Extension performance depends directly upon its capability.

Effectiveness is defined by a handbook on productivity management as "the degree to which goals are attained". Agricultural extension has many goals such as social goals (e.g., farmer welfare) and economic goals. Among these, operational goals (e.g., physical and financial targets) are of special significance because their attainment makes realisation of other goals possible.

Efficiency in extension is usually measured by the rates at which farmers adopt recommended practices. Adoption rates of varying degrees of complexity can be conceived.

Impact in extension can be measured by a simple indicator, like yield of a crop per hectare, or by constructing simple productivity indices. Such indicators provide ultimate tests for the success of extension programmes.

EXTENSION MONITORING INDICATORS

Extension monitoring indicators can be grouped into two categories:

(1) extension capability indicators, and

(2) extension performance indicators.

Both should be generated by the monitoring unit.

Extension Capability Indicators

Extension capability indicators must be monitored regularly not only to know the status of extension's capability at a certain point in time, but also to determine changes in it over time. These indicators should be calculated annually. They involve only desk work because they are based on in-house data.

Extension Performance Indicators

Extension performance indicators reflect extension's operational and technical efficiency. They can be grouped into two categories.

Extension Effectiveness Indicators

These can again be grouped into two subcategories:

(1) single indicators and

(2) unitary or composite indicators.

By definition, a single indicator will reflect an aspect of extension performance, while a unitary or composite indicator will reflect two or more aspects of extension performance. It may be useful to construct a unitary or composite indicator to provide a consolidated view of extension effectiveness to management, because management is often interested in having an overall view of extension effectiveness. These indicators reflect extension's cooperational efficiency.

Extension Efficiency Indicators. These indicators are based on adoption rates of recommended practices and reflect extension's technical efficiency.

Extension performance indicators, as suggested here, should be calculated separately for contact farmers, other (noncontact) farmers, and all (contact and noncontact) farmers.

Extension evaluation indicators

Yield and productivity indices occupy the central position in these groups of indicators. They are calculated on the basis of crop-cutting experiments which are conductebby provincial revenue or economics and statistics departments.

Action Plan and Chain of Events

Monitoring requires collection of data and their analysis. For this, an action plan in a monthly, quarterly, and annual timeframe is essential. Data collections will require choosing appropriate methods, which will depend upon the monitoring unit's time and physical and financial resources such as computerisation and trained technical personnel for collecting and analysing data.

Monitoring methodologies can be perceived as a chain of distinguishable and sequential events comprising

(1) planning and design of the study,

(2) desk research,

(3) selection of methods,

(4) data collection and analysis,

(5) report writing,

(6) report presentation, and

(7) follow-up action.

An array of methodologies is now available, ranging from casual, informal interactions to highly structured sample surveys, including emerging methodologies of rapid appraisals. These methodologies can be grouped into two categories: informal and formal.

Informal methods

Informal methods include participant observation, case studies, key informants, individual interviews group interviews or discussions, oral testimonoral and life histories, longitudinal studies, cross-sectional studies, interdisciplinary terms, sondeo or reconnaissance survey, diagnostic studies and rapid rural appraisal.

Formal methods

Formal methods include using the population and agricultural census; sample surveys such as random sampling, including simple, systematic, stratified, cluster, and multistage; nonrandom sampling, including purposive, quota, and accidental; and special studies. Among these, the sample survey occupies a unique place. A sample, as a representative part of the population, enables us to draw reasonable inferences about the characteristics of the population. As such, the sample survey will constitute the "handover" of monitoring methodologies. Other methods, notably the emerging rapid rural appraisal, can be used to advantage for exploratory, diagnostic, complementary, and supplementary purposes for decision making.

REFERENCES

Hortan, D., Peterson, W., & Ballantyne, P. 1993.. M & E principles and concepts p. 5-19.. In D. Hortan et al. Eds.., *Monitoring and evaluating agricultural research: A sourcebook.* Oxon, UK: CAB International.

Israel, A. 1987.. *Institutional development: Incentives to performance.* Baltimore and London: Johns Hopkins University Press published for the World Bank..

Kumar, K. Ed... 1993.. An overview of rapid appraisal methods p. 1-22.. In K. Kumar Ed.., *Rapid appraisal methods.* Washington, DC: World Bank.

Lee, J. W. 1990.. *More than accountability: Evaluating agricultural extension programmes.* TRDC Publication No. 172. Amidale, Australia: University of New England, Rural Development Centre.

Misra, D. C. 1994.. *Agricultural extension effectiveness in India mimeo..* University of Oxford, Queen Elizabeth House, International Development Centre.

Murphy, J. 1993.. Good enough, soon enough. *The Rural Extension Bulletin, 1,* 27-33.

Murphy, J., & Marchant, T. J. 1988.. *Monitoring and evaluation in extension agencies.* World Bank Technical Paper No. 79. Washington, DC: World Bank.

Nichols, P. 1991.. *Social survey methods: Afield guide for development workers.* Development Guidelines No. 6. Oxford: Oxfam.

Pétry, F. 1983.. Designing a monitoring and evaluation system p. 20-49.. In E. Clayton & F. Pétry Eds.., *Monitoring systems for agricultural and rural development projects* vol. 1.. Rome: FAO.

Pratt, B., & Loizos, P. 1992.. *Choosing research methods: Data collection for development workers.* Oxford: Oxfam.

17

Evaluation of Extension Programmes

Although extension educators, funders, and administrators are in favour of evaluating extension programmes, honesty requires us to admit that most of us are not overly enthusiastic about undertaking it. There are many reasons for resistance. However, if we are to provide evaluation leadership, we must recognise forms of organisational and personal resistances, or these will sabotage our evaluation efforts.

It is not uncommon to use the occasion of evaluation to mask our insecurity with distrust or criticism of others' intentions, motivations, competency, or adequacy of effort. We are sometimes caught in a bind between our own reluctance to ask ourselves hard evaluation questions and our resentment of outsiders asking questions. The second form of resistance is fear, particularly the fear of change that evaluation might precipitate. It is the nature of organisations to be self-protective and defensive.

Evaluation and organisational comfort appear to be somewhat incompatible because evaluation challenges organisations to change. We all resist change to some extent. Although evaluation is not revolutionary, it is a handmaiden to gradual change, and we have to recognise that our reluctance to participate in evaluation is partially a reluctance to embrace needed change. Another resistance to evaluation comes from our need to avoid embarrassment about potential bad news. We need to know the quality of our efforts, but we also have a fear of finding out the truth about our achievements, particularly if we lack confidence. Most of us avoid tests for the same reason. We need to face up to our personal and organisational ambiguity regarding our need to know and our fear of knowing. Still another real resistance results from the fact that evaluation is often an additional task to an already impossible workload.

Those whose job description includes evaluation may need only to be reminded of this. However, potential benefits may have to be discussed and identified if collaboration is to become a reality among those who do not have a formal professional responsibility for under-taking it. Benefits include recognition for achievements, opportunities to improve practice, establishment of accountability, and learning new lessons about our efforts.

CREATING POSITIVE EVALUATION METAPHORS

The term "evaluation" can be threatening. If evaluations are to be positively embraced by extension organisations, it may be necessary for those initiating them to be positive in the language they use to describe the proposed effort. Negative metaphors that are sometimes used for evaluation, such as "under the gun," "a mine field," "an invasion of inspectors," "last rites before hanging," and "defend the fortress" reflect past negative experiences with evaluation. Why not create more positive metaphors for the evaluation task? For instance, we could describe evaluation in medical metaphors such as "complete programme physical" to identify physical soundness and potential programme health risks.

The term "auto-diagnostico" has been used in the Dominican Republic and "self-strengthening" in Sri Lanka. Or we could use a business metaphor, such as "taking stock" of profits and liabilities. We could use sports metaphors, such as "watching video tapes" of yesterday's game to improve our "game plan." We can use "mirrors" to see ourselves as others see us. We could consider evaluation to be a "learning process" or a "sabbath" day of reflection, renewal, gratitude, and celebration. Being self-conscious and deliberate about the language that is used regarding evaluation may help reduce unnecessary resistance to the effort and may build positive support.

Exploding Myths about Evaluation

Several evaluation myths have often discouraged extension managers from engaging in useful evaluation.

Myth #1: Evaluate only when mandated. Many funded programmes require evaluation as a form of accountability. However, it is a myth that evaluation should occur only if it is mandated. On the contrary, evaluations that are self-initiated are more likely to be taken seriously for immediate programme improvement. Programmes become responsible and excellent just as often through self-initiated evaluation.

Myth #2: Evaluation is an add-on. It is a myth that evaluation is an add-on activity or at most a pretest with a posttest. It is most meaningful when it is integrated with decision making at every stage of programme planning and operation.

Myth #3: Evaluation is an activity for experts. It is a myth that evaluation should be undertaken only by technical experts. Yes, complex methods can be used; however, systematic evaluation can be undertaken by inexperienced managers, and specialists and educators themselves can be helped to critique their own work.

Myth #4: Outside evaluators are best. It also is a myth that evaluation should be done only by external, outside, objective evaluators. Yes, external evaluators are often useful in challenging insiders to address what they have overlooked because of their nearsightedness. However, internal, self-initiated, and subjectively oriented evaluations also can be rigorous and valuable. In fact, because they often are participatory in generating, analysing, and interpreting data, they may result in greater acceptability of the findings and recommendations.

Myth #5: There is one best evaluation approach. Still another myth is that there is one best way to conduct an extension evaluation. Some approaches are probably better than others for addressing particular types of questions or concerns. However, the many types of evaluation approaches have their own strengths and limitations.

Myth #6: Quantitative data are best. A mixed-methods approach combining qualitative and quantitative methods can lead to better understanding and appreciation of phenomena under evaluation and provide triangulation, convergence, and corroboration of results from different methods. Qualitative methods are best for understanding the nature of something, while quantitative methods help in appreciating its extent. If we do not know the nature of something, we should conduct qualitative studies. After measuring something, we may still need to use qualitative methods to learn about variations and unique forms. The use of mixed methods can increase user responsiveness to evaluation information. We should ask which combinations of methods are best for answering which evaluation questions, rather than deciding on a method and then forcing or changing our questions to fit that method.

MAJOR ELEMENTS IN EVALUATION

There are at least five major elements in most evaluations: (1) focus questions, (2) objects or events to be evaluated, (3) data or evidence, (4) analysis and interpretation using judgement perspectives, and (5) judgements, conclusions, or findings. Purposes and approaches or models may vary, but these elements will be present in one form or another.

Selecting evaluation purposes:

There are many purposes for undertaking evaluations in any particular situation. It cannot be assumed that all stakeholders (farmers, extension staff, administrators, funders) share common purposes. Although people may never completely agree about the purposes, one of the tasks at the outset is to seek clarity about evaluation purposes with major stakeholders. Appreciating different purposes at the outset can reduce conflicts and disappointments down the road. Consider the following purposes.

Pseudo self-serving purposes

Since organisations, including extension systems, have a self-serving tendency, it is not unreasonable to expect that some staff members, especially those in the highest places, may want a pseudo evaluation that will postpone, buy time, or avoid threatening change. In these cases, evaluators are not taken seriously, and the evaluation becomes a meaningless political diversion. In other cases, some members of organisations want evaluations as excuses for evading or avoiding administrative responsibility or to provide a scapegoat for criticism. Evaluations that are undertaken only to make the programme look good or to make someone or some aspect of a programme look bad are pseudo and illegitimate.

Enhance accountability purposes

It is quite common for external donors to expect that evaluation will provide accountability through evidence of impact, or to document cost-benefits, or to measure efficiency-effectiveness. In some cases, this evaluative evidence is considered in decisions to continue the programme; or to propose change, expansion, or reduction of a programme; or to change a policy, organisational structure, philosophy, or design. The potential for negative findings and the threat of discontinuing funding has led to "hiding the mistake," a dysfunctional practice. However, evaluations rarely provide a single basis for political decisions. They often are used by funders, administrators, or policy makers to justify their decisions even when the evidence of benefits is weak.

Improve performance purposes

This purpose of evaluation is sometimes called "formative" because the results are intended to help improve the programme during its formative stages. This is in contrast to "summative evaluations" when the purpose is to sum up or summarise the accomplishments at a point in time. When evaluations are to improve programmes, lessons learned about strengths and limitations of the programme are mined from the data so that changes can be made immediately. Sometimes the intent is to discover new approaches and alternatives or to adjust the programme to changing situations or client groups. Evaluation also is used to understand multiple reasons for apparent failure or to improve the management or operation of a programme.

Social learning and communication purposes

Sometimes evaluations are intended to stimulate political dialogue or to resolve political conflicts intelligently. For example, an evaluation of extension in a country could provide an opportunity to debate the need to hire more women agents to respond to an increase of women in small-scale agriculture or to extend the extension network to subsistence farmers not being served. Often the most significant contribution of an evaluation is the creation of new expectations, new organisational arrangements, new linkages, and new purposes and goals. Evaluation may give visibility to a good idea and new language that can communicate new ways of viewing extension to others who also may want to share an experiment.

Evaluation purposes tend to vary, depending upon where one stands within a system like extension. External funders often want an accountability purpose, while field staff are more likely to favour a programme improvement purpose. Policy makers and programme administrators can often appreciate an evaluation that contributes to new ways of thinking about extension or new forms of extension. Farmers want an evaluation to improve the benefits they may receive from extension staff.

Recognising the Politics of Evaluation

Evaluations are never value neutral. Political implications are always present because stakeholders have interests to defend. In this sense, all evaluations are political. Therefore, it is very important to know the values, expectations, and interests of the stakeholders at the outset. Failure to understand these interests may make the findings irrelevant or may cause unnecessary conflict and rejection of the findings. Knowing the interests also helps in the focus and design of the evaluation. Values, expectations, and interests are reflected in stakeholders' visions regarding what "good" looks like in extension. Some may favour an extension system that empowers the poorest of farmers in their struggle for more land and a larger share of market benefits.

Multinational agribusinesses may view the success of extension as increasing productivity for export. What a "good" or "excellent" extension system should look like will vary according to these different values and criteria. What different funders want may differ from what various groups of farmers may want in an extension system. Assumptions of stakeholders also vary regarding the relationship of extension education to larger social and political visions.

For some, the extension system is intended to reproduce existing power relationships between government and farmers. For others the extension system is viewed as an instrument to transform those relationship. For some, extension is primarily a technology transfer system from researchers to farmers. For others, extension is a communication network among farmers, researchers, credit institutions, market organisers, consumers, and government policy makers. These expectations, values, and interests are reflected in the criteria that various stakeholders put forth as central for judging extension, regardless of the stated objectives in programme documents.

External Evaluator	Internal Evaluator
External Criteria	External Criteria
External Evaluator	Internal Evaluator
Internal Criteria	Internal Criteria

Figure 1. Alternative locus of evaluators and criteria for planning extension evaluations.

Political interests are usually behind the debate over external versus internal evaluations. Some external stakeholders favour external evaluators because they want an evaluation to serve the direction of change, and they suspect that internal evaluators will be self-protective and not face up to the necessary reforms. Local programme staff sometimes fear external evaluators who impose criteria or external visions through evaluation judgements without understanding the situation on the front lines. They do not want to be evaluated against unreasonable external criteria.

Education, including extension education, also is about power, because one way of increasing power is through increasing knowledge. Extension is political in that it can either reproduce existing access to knowledge or privilege or provide a redistribution of power in a society through increasing knowledge access.

Selecting Alternative Approaches and Models

What are several alternative approaches to programme evaluation? Choosing among these approaches is important because for each there are different assumptions about what data to collect, how to collect them, and how to make judgements about success. The following seven major approaches will provide a sufficient choice for most extension evaluation situations: (1) expert model, (2) goal-free model, (3) attainment of objectives model, (4) management decision model, (5) naturalistic model, (6) experimental model, and (7) participatory evaluation model.

Major Models for Programme Evaluation

Expert Model. This approach relies on expert judgement. Usually, documentation is prepared in advance of experts' visits. The experts then interview, analyse documents, and make judgements using their own judgement perspectives or those set as standards by the outside organisations or stakeholders. Typically this type of evaluation brings in a team of experts from FAO or extension systems from several countries to make judgements and comparisons regarding strengths and limitations.

Goal-free model. This approach assumes that outside evaluators do not know, or need to know, what the programme has intended to accomplish, but that it is the task of the evaluators to uncover what is actually happening relative to farmers' interests regardless of stated goals and intentions. The focus point is to identify environmental and farming conditions and then to compare these needs with what people are actually experiencing as a result of the extension programme. The gap is then viewed as a starting point for making changes in the programme. An example is an evaluation that describes conditions of indigenous farming groups cultivating fragile hillside soils and comparing these conditions with access to and appropriate content of knowledge from existing extension services. This approach relies heavily on open-ended interviewing and observation by persons who do not have a vested interest in the programme.

Attainment of objectives model. This approach assumes that the success of a programme can be determined by measuring a programme's outcomes against its own goals and objectives. This type of evaluation begins with clarifying measurable objectives and then gathering data that validate the extent to which these objectives have been met. For this model to be credible, an essential feature should be added, namely, the evaluation of the appropriateness of goals and objectives, given the circumstances and needs of farmers.

An extension system may have adequately met its objectives of increasing production of maize among large landholders, but at the same time it may have neglected to question its lack of commitment to small landholders or tenant farmers. If an attainment of objectives evaluation is anticipated, programmes are often tempted to set goals quite low so that outcomes will be met easily, thus appearing to be successful while ignoring major challenges. This model also has a "black box" limitation in that it tends to ignore the extension process, thereby failing to provide explanations for outcomes.

Management decision model. The purpose of this model is to provide relevant information as a management tool to decision makers. It assumes that evaluation should be geared to decisions during programme initiation and operation stages to make results more relevant at each particular stage. Participation of stakeholders is central to the process because evaluation should serve their decisions. Sometimes cost effectiveness and operations monitoring are included.

One limitation of this model is the tendency for the decisions of major stakeholders to be viewed as more important than those of various types of farmers, especially women in agriculture who may not benefit directly from such an evaluation unless care is taken. Naturalistic Model. This model assumes that a programme is a natural experiment and that the purpose of evaluation is to understand how the programme is operating in its natural environment. There is an assumption that programmes are negotiated realities among the significant stakeholders and that evaluation serves this value-laden negotiation. Data should be collected and analysed from multiple perspectives.

The outcome of the evaluation is dialogue concerning disagreements about objectives, expectations, problems, opportunities, policies, procedures, and suggested changes in methods or activities. Many positive collaborative changes can be made through this model of evaluation if conflict resolution skills are combined with evaluation. Another purpose of this model is to diagnose or to identify the causes for certain behaviour on the part of some farmers, agency staff, or other development actors.

Experimental model. The purpose of this approach is to determine whether changes in programme outcomes (learning accomplishments) were due to the contributions of the programme and not just to life's experiences or from other influences. This model asks the question, "Were differences in sustain-able agriculture practice attributable to the programme?" The simplest way to determine causality between the programme inputs and comparable groups, a group that received the educational treatment and a group that did not. This means that programme accessibility, at least during the experiment, programme consequences is to compare at least two is withheld from those learners who serve as a control group. Because of the nature of human subjects, the ethics of withholding educational services, and the difficulty of controlling for external influences, it is extremely difficult and costly to operationalise this model. It is recommended that this model be used only when major changes are expected or when a major failure is anticipated in pilot efforts where causal claims are central to making major programme investments.

Participatory evaluation model. The purpose of this model is for extension educators and farmers themselves to initiate a critical reflection process focussed on their own activities. This is done through identifying a persistent major situation such as extension's neglect of women in agriculture; subject it to critical reflection, underlying assumptions, habits of mind, and cause and effect expectations; and then after creating new assumptions, change practices and validate or invalidate the results. The model assumes a democratic participatory process along with autonomy on the part of educators and learners at the local level. This is a form of what is usually called "participatory action research."

Benefits and limitations of participation in evaluation

Why use participatory approaches to evaluation? Involving people who are on the receiving end is likely to assure the most efficient allocation of scarce resources and the early identification of ineffective or wasteful use of resources. People on the receiving end are ultimately the best judges of impact, whether benefits have been produced or not. Being included in planning, implementation, and evaluation will show farmers that they are regarded as responsible, capable individuals and not simply passive "beneficiaries" or a "target group."

Participatory evaluation is self-educating. It can encourage the development of human capacities among farmers. Without participation in evaluation, sustainable agriculture is unlikely. Participatory evaluation also can decrease the paternalistic, directive, impatient, or insensitive relationships among officials, technicians, and farmers by improving staff attitudes toward working with persons having less status and education. Although an initial cost of time is required, participation in evaluation can speed implementation when participants take greater ownership of efforts.

Farmers' lack of experience is sometimes said to be an impediment to participatory evaluation, but this need not be an obstacle if the process whereby people gain experience can be planned for and invested in. Lack of resources and organisational skills can be overcome. Power differentials, social stratification and cleavages, and personal conflicts do present limitations. However, these obstacles can be reduced by outside actors who can provide spaces for subgroups to express their views for the sake of broader community interests.

Officials or NGO staff themselves may present the most formidable obstacle by their paternalism and preoccupation with control. They often feel threatened by people participation. Organisations tend to replicate in their environments the same attitudes, values, and social relations they exhibit internally. Practising participatory evaluation internally will provide a model for participatory evaluation that includes farmers. Positive experiences with participatory evaluation can help overcome staff anxiety.

How much participation should the farmers have in evaluating extension programmes and their own learning? Farmers, in many extension programmes, have participated in evaluation of extension through farmer associations and committees. However, as Arnstein has pointed out in a much quoted article regarding a ladder of participation, there are various levels of involvement and participation. In her model, the bottom rungs provide for merely token participation, while the top rungs allow for participant control.

Focusing the Evaluation Effort

What should be the focus of programme evaluation? This question raises the spectre of evaluating everything, which is an impossibility. Choices and priorities among many possible questions have to be made. A full range of possible focusses can be represented at least in part by the levels represented in Table 1, which shows eight major areas of focus for programme evaluation:

(1) inputs,

(2) activities,

(3) participation,

(4) reactions,

(5) individual change,

(6) organisational change,

(7) community change, and

(8) national change.

Evaluations rarely cover all items of these areas. Most limit themselves to a combination of items that serve the evaluation model or the concerns of stakeholders. Narrowing the focus usually begins during planning with the major stakeholders in a programme effort. Interviews with these stakeholders are usually conducted at the outset of evaluation efforts, often during the programme-planning process, to identify the programme model or approach and to determine the questions that are central. Stakeholders often do not agree, so priorities, methods, and costs must be negotiated so that the task can be "reality based" and "doable." The key here is to determine the decisions that stakeholders intend to make based on the evaluation findings so that immediate use will be made of the information generated. Evaluation use is not based on the quantity of data, but on its timeliness and relevance to decisions and purpose. The seven evaluation models described above tend to focus on the different levels shown in Table 1.

Table 1 : Focus for programme evaluation.

1. *National impacts:* Political stability, economic fairness, agricultural environmental sustainability
2. *Community change:* Changes in administration of justice; health, welfare, and quality of life; fairness in the marketplace; change in human rights, status of women; change in economic and social indicators for poor; change of indicators of sustainable agriculture and natural resource management; change in communication patterns and access to education and news; public opinion change; fairer distribution of land and other resources; improved interorganisation relations; evidence of conflict resolution; and cultural practice change
3. *Organisation change:* Group operation and management; economic performance; technical operation and management; financial operation; group institutionalisation and self-reliance, new groups of farmers included, new organisation linkages; change in staff performance, new service delivery, new methods used, additional facilities and equipment; cost-benefits improved; new philosophy, purposes, and goals; improved organisational culture
4. *Individual change:* Changes in knowledge, attitudes, skills; sustainable agricultural practice; change in aspirations, self-image, perspectives; expenditure of effort and money; use of methods, services; invention of appropriate technology; increased production or use of tools; compliance with or opposition to public policy; patterns of communication, career directions, and family relationships
5. *Reactions:* Testimonials; reactions to the relevance of content; appropriateness of technology, helpfulness, perceived value of educational experience; reputation of the extension provider
6. *Participation:* Farmer access to extension services by social class, gender, and ethnic groups; intensity of face-to-face contacts; extent of media-assisted contacts; type of participation (volunteering, planning, recruiting, learning, experimenting, evaluating); indicators of commitment (attendance, continuity, frequency)

7. *Activities:* Participatory rural appraisal; planning; local knowledge documentation; farmer experimentation; farmer-to-farmer knowledge sharing; farm tours; farmer organising; master farmer leadership training; farm demonstrations; exhibitions and fairs; residential workshops; marketing analysis; farm policy education

8. *Inputs-resources:* Organisational sponsorship and networks; funds; organisational design, facilities, equipment; philosophy, mission, goals, objectives; staff, resource people, volunteers; local and external research knowledge and relevance; cultural, economic, and political context.

For in-stance, the expert model most often focusses on data from inputs, activities, and participation, while the goal-free model tends to focus on individual change, organisation change, or community change, ignoring the inputs and activities. The attainment of objectives model usually compares the philosophy, goals, and objectives of inputs to the extent of individual or organisational change outcomes.

The naturalistic model emphasizes understanding activities, participation, and reactions as processes that occur within cultural, economic, and political contexts. The experimental model emphasizes causal relationships between inputs and individual or organisational change. The participatory evaluation model emphasizes activities and their relationship to benefits and values to farmers. It also emphasizes participation of farmers themselves in planning the focus, data collection, interpretation, and implementation of action that emerges from the evaluation process.

Selecting Methods for Programme Evaluation

Which methods are right for the task? The basic rule here is that the selection of methods follows the selection of focus, not the other way around. Each evaluation question must be examined in relationship to what would constitute evidence for answering it. The following brief descriptions of data collection methods, although by no means exhaustive, can be used as a toolkit for a variety of circumstances. The list includes document analysis, observations, interviews, surveys, focus group committees, field visits and tours, village drama and role plays, maps, case studies, field trial documentation by farmers, remote sensing, and aerial photography.

Document analysis. Examples include minutes of meetings, correspondence, budget records, workshop notes, participant papers, and newspaper reports, to name a few. These can be treated as data, analysed for content, and summarised in relation to questions, including extent of inputs into the programme; levels of participation, nature of goals and activities, and themes regarding problems, concerns, expectations, and new directions. Themes from documents can be a credible source of information. These documents usually do not reveal participants' motivations or subjective experiences. However, documents often reveal difficulties of programme operation.

Observations. Observers can be outsiders or persons who are involved in learning activities. Observers are usually given a short list of items that may include extent of participation and personal interaction, nonverbal indicators of interest or inattention, leadership roles, performance levels, and conflict indicators, to name a few. Both qualitative and quantitative data can be collected.

Findings can be reported to the learners as a whole to start a reflective process about what may need to be changed, or they can be used as evidence of successful methods or learning outcomes. Observations of process and outcomes can be recorded by video or photo documentation. These data are very powerful graphic ways of communicating the nature of a programme and its outcomes to sponsors as well as to potential learners. Video records of learners' local knowledge also can be used to help them reflect on strengths and limitations of knowledge.

Interviews. Interviews are probably the most widely used method for programme evaluation, including the evaluation of learning. Interviews with key informants and representative farmers are suited to indepth explorations of issues. If the questions are standardised, responses can be tabulated numerically to indicate item strength. If questions are open-ended, indepth unique responses can be generated, which in turn can provide information regarding reasons why the activities are viewed differently by diverse groups of participants.

Individual oral histories can inform patterns of practice and the use of extension resources. Storytelling is one of the oldest forms of human social life. Evaluative judgements are usually embedded in stories. Interviews, oral histories, and Storytelling with farmers who are not served is essential to a full picture of extension. Illiterate members of the community can participate fully through interviews, oral histories, and Storytelling, demonstrating that illiteracy does not rule out participatory evaluation.

Group interviews (sometimes called focus groups) can be formed according to geographic locations or by farm type to discuss specific evaluation questions. Sometimes the groups may already exist. At other times new groups may be formed just for the evaluation. The purpose is not only to generate judgements using agreed upon criteria, but also to uncover unanticipated outcomes, applications, opportunities, and problems to inform future extension efforts.

A community can reconstruct its history, chronology of events, crises, turning points, accomplishments, and so forth. Chalkboards, pictures, and drawings can represent milestones or decisions. The art of conducting group interviews can be learned. Many community members have a talent for this and are more likely than agency staff to evoke authentic responses. Subgroups are often necessary to listen to voices that are unlikely to be heard in groups dominated by those with status.

Surveys. The survey is a more standardised form of data collection that incorporates a prepared questionnaire. In most parts of the world, surveys of farmers in rural areas must be conducted by interview using interviewers who know the territory, language, and culture of potential respondents. Sometimes surveys can be administered at meetings or public gatherings; however, responses have to be treated as an "opportunity sample," rather than as a "random sample," and therefore generalisation of the findings to a larger population is limited.

Surveys are often used to evaluate the extent of practice, estimations of production yields, preferences for appropriate technology, and expectations regarding the future. They are best used with homogeneous populations rather than with quite diverse populations because standardised questionnaires are less likely to be sensitive to diversity. Evaluation of unique practices and adaptations of diverse farmer types is best done through interviews and observations.

Field visits and tours. There is no real substitute for field visits and tours to provide authenticity and reality to the conditions, limitations, and impacts of extension programmes. Evaluation teams composed of local farmers, extension staff, administrators, funders, and external evaluators provide balance and interactive learning regarding different perspectives. Team members can undertake observations and interviews and learn from each other about their findings during travel. Assignments can be distributed so that special knowledge regarding specific aspects of situations can be gathered.

Some team members can focus on economic, social, and cultural aspects, while others focus on technical aspects. Comparing data, analysis and reflection on findings, and insights following field visits can generate a more holistic, balanced evaluation.

Documentation of farmer demonstrations. Farm visits, which have been used frequently by extension staff for teaching and the transfer of technology, also can be used for evaluation of appropriate technology. Farmers and local leaders can be taught to conduct their own field trials, thus fostering the pride and dignity of local people who can transfer appropriate technologies through their local languages. When farmers choose their own focus of inquiry, collect and analyse their own data, and come to their own evaluative judgement, they are more likely to adopt and pass on relevant and effective appropriate technologies. World Neighbors NGOs in Bolivia and Peru have developed a standardised means by which farmers engage in site-specific experimental documentation and reporting of crop yield comparisons, using simple calculators.

Village drama and role plays. Asking farmers in farmer association meetings or in village gatherings to create a drama or role play that describes the interaction process of extension staff with the village on a specific practice will reveal a variety of evaluative data on social relationships, relevance of extension knowledge to local knowledge, and historical events that have affected the solutions to farmer problems.

Maps. The generation of maps can provide a basis for making judgements about access to extension resources by showing where farmer contacts have been made. Maps also can be made to show the location of sustainable agricultural practices. These facts can be used to evaluate the scope and effectiveness of extension efforts on practices in a region, watershed, or geographic area. When access maps are overlaid with social ranking maps, judgements can be made regarding benefits by social class. Concept maps created collaboratively among farmers and extension staff can provide explanations regarding success and failure of specific programme efforts. Reflections on these maps, which can be drawn on chalkboards or in the sand, can reveal contradictions in underlying assumptions and expectations, which in turn can lead to new experiments.

Maps showing before-and-after photographic and remote sensing images can be the basis for evaluative regarding sustainability of farmer and extension staff practices. This is especially relevant to desertification, deforestation, soil erosion, and status of wetlands and wildlife habitat. These data can help direct extension activity to environments that are most at risk, as well as provide evidence regarding extension's positive impacts from collaboration with farmers on sustainable agricultural and natural resource practices.

Case studies. In order to understand motivation of farmers or potential extension contacts, case studies of specific types of farmers or farming practices can be undertaken. Comparisons between farmers who have used extension technologies and those who have not are common types of case studies. The typology may be based on geographic regions, soil types, and cultural, age, gender, and economic differences. Case studies are best constructed through repeated interviews over time and often include, in addition to self-report, data from persons who know the subjects well. Oral histories, logs, and journals can also contribute to case study data if farmers collaborate in producing these case studies.

SELECTING METHODS FOR EVALUATION OF TEACHING AND LEARNING

We have briefly considered methods for programme evaluation at the macro level. Let us now consider the evaluation of learning at the micro level from the perspective of the two most important stakeholders, the extension educators and the farmers or programme volunteers as learners in specific educational events. When learning is evaluated, there are many questions to answer.

Central of course is how learners experience the learning process and what they actually learn, their knowledge, attitudes, skills, and aspirations, and their behaviour change. An evaluation also can focus on the extension educator as learner and the content, processes, and resources that are used. Because learning is always a social phenomenon, an evaluation can focus on the social environment, organisational context, and the relevance of language, culture, and sometimes public policy to learning. These underlying cultural assumptions often explain resistance to learning, as well as the way learning either reproduces existing racial, gender, and economic power relationships or challenges these relationships. Not all evaluations include all of these questions. Educators tend to focus on questions that serve their own perspectives. Learners, including farmers, likewise may be interested in questions that serve their perspectives. Often adult learners are eager to reflect critically on their past and present learning contexts in order to overcome socially constructed deterrents to their learning. When farmers consider evaluation of their own learning, they may ask themselves a broad range of questions.

Perspective of Extension Educators

When extension educators consider evaluation of learning, they usually want to know how the learners perceive the process of learning, especially how they, as educators, have been helpful to the learning process. Asking learners to reflect on educator behaviours also encourages critical reflection on their own learning. The purpose of listing all of these questions from both the learners' and educators' perspectives is not that one should ask them all, but to stimulate about which are essential items for a specific evaluation effort. The methods that follow can be used to gather evidence of learning through relationships.

Documentation of local and external knowledge. A basic principle of adult education is that learning should begin with prior knowledge. How can we appreciate what has been learned if educators do not know what participants already know? For centuries, farmers have passed on their indigenous or local knowledge. Extension educators can assist farmers in documenting this knowledge either in written, photographic, audio, or video forms. This is essential, since evaluation of new learning should be compared with what farmers already know. Also, farmer acknowledgment of the limitations of their local knowledge can form the basis for collaborative inquiry and the linkage of external knowledge with their local knowledge.

Rating scales and checklists. Educators can use these in checking the performance of learners. Learners can use these to check the performance of educators. They can be administered in group or field settings and can be easily revised. Learners can use them to judge their own performance, current knowledge, or educational expectations. Rating scales and checklists are not very useful in measuring attitudes or consequences of performance.

Feedback committees. When residential training for extension staff and farmers takes place, participants can elect a feedback committee to provide evaluation observations to the leadership during the event. The feedback committee should be open to any complaints participants may have about the event, ranging from relevance of content, adequacy of facilities, or effectiveness of leadership, to the involvement of learners in activities. The committee can bring items to the attention of the instructor or to the group as a whole through written or oral form.

Group discussion assessment. This method can be incorporated into an ongoing group or meeting. It is relatively efficient in terms of costs and time use.

Peer review panels. Farmers can become involved in evaluating one another's work through peer review panels. Panels can be taught to use standards and rating scales. Their evaluative judgements can be made with or without identification of reviewers. When peer review panels are used, it is important to establish a positive climate of constructive criticism.

End-of-event analysis. This can be done in several ways. The most frequently used is an evaluation form that is administered at the end of workshop sessions. Another way is to have these forms distributed, collected, and summarised by a feedback committee.

Testimonials and stories. Testimonials and stories can provide subjective records of educational experiences and activities from the perspective of the learners. They are a form of results data and can qualitatively describe the nature and process of educational change. These stories also can be easily understood by others outside the programme as illustrations of types of outcomes and can lead to ideas for future programming. Disadvantages of this method include social desirability bias, nongeneralisability beyond the person giving the testimony, and difficulty in determining what happened as a result of the programme versus other influences on the person. Stories can be either written by the learners or created as a result of an audio-taped interview.

Pretests, Posttests, and Quizzes. In spite of the negative attitudes associated with tests and quizzes, they can be useful for diagnosing learner proficiencies, documenting prior knowledge, projecting learning achievements, and understanding learner attitudes.

Interpreting Findings and Data

In some evaluations, the findings consist of narratives that describe the history of events. In other cases, evaluation findings are descriptions of the way things are without explanations or judgements. In still other cases, evaluations seek to answer questions of cause and effect or the relationships between methods and outcomes.

Standard-referenced Judgement perspective

Findings regarding a particular extension programme can be compared with examples of excellent achievement by using agreed upon criteria established by experts as "state of the art." For instance, such experts could describe excellence as (1) promotion of sustain-able agricultural practice; (2) recognition of local knowledge and cultural practices; (3) high participation of farmers in appropriate technology generation; (4) high inclusion of farmers from all economic, social, and cultural groups; (5) access of women farmers to extension; and (6) promotion of justice in agricultural policies.

Cohort-referenced judgement perspective

Findings from an extension programme evaluation can be compared with similar programmes. For this judgement, we ask how a particular programme or aspect of a programme would be ranked with extension programmes elsewhere. External evaluators are most prepared to do this since they may be knowledgeable about extension practices and impacts elsewhere. This approach is limited because few places are truly similar in extension resources, political and economic contexts, and other handicaps that can affect programme outcomes. However, a failure to make judgements using external cohort comparisons may result in a false sense of achievement and self-satisfaction. Such programmes, when compared, may be mediocre, outdated, poorly conceptualised, and wasteful while being considered excellent or acceptable by local practitioners and their funders.

Difficulty-referenced judgement perspective

This perspective takes into consideration the difficulty of what is being attempted when making judgements regarding programme achievement. For example, an extension programme that is addressing sustainable agriculture in highland areas with few staff members in the midst of revolutionary unrest must be given extra credit for achievements under these difficult circumstances compared with programmes that have a large staff who are helping farmers with relatively rich soil increase their crop production. Difficulty-referenced judgement is the basis for giving higher scores for more difficult dives in Olympic competition. Achievement must be judged relative to difficulty and conditions.

Progress-referenced judgement perspective

This perspective on the interpretation of findings gives credit and recognition to progress from past to present. Before-and-after descriptions are essential to making these judgements. Those who emphasize "base line" data usually want to use it to make progress-referenced judgements.

Alternative-referenced judgement perspective

This perspective considers present descriptions of an extension programme in comparison with what could have been accomplished with the same resources used in alternative ways and places or with different people. This perspective asks, "How else could these resources have been spent with better or different results?" For example, would sustainable agricultural and environmental practices be more evident today if funds had been given directly to village groups or farmers' associations for conducting their own participatory action research? Is the present farm visit system better than a distance education system using radio and audio tapes? What would have been the results if all areas of extension programmes had worked more directly with women through their traditional organisations? Sometimes answers to these questions can be generated by projecting results based on pilot efforts or by alternative approaches that have already been tried in limited form in parts of an extension system. These judgement perspectives also are helpful in evaluation of learning by individual participants themselves. Learners can judge their learning in comparison with expert standards, performance by their peers, their learning handicaps and external difficulties, their own progress from past to present, and what they would have found more important to have learned or done with their time. Often a combination of these judgements provides a balanced evaluation.

All too often decisions are made to undertake an extensive evaluation without counting the cost and without people agreeing to do the work. These efforts are doomed to failure. Negotiating what is possible, given resource limitations and reasonable time expectations, is an essential collaborative task if people at all levels are to own the evaluation and act on the findings. Involving learners and educators together in critically reflecting on and interpreting the evidence will go a long way toward guaranteeing that evaluation itself is a major learning process. Typically, a single report provided to stakeholders neglects the interests of many other participant groups. One solution is to consider the findings, interpretations, and judgements as constituting a pool from which a variety of reports and styles of reporting can be fashioned to serve specific purposes and different users. Forms of reporting can include formal written reports, written executive summaries, letters to individuals and organisations, exhibits and pictorial displays, video narratives, news reports, news conferences, and public meetings. Stakeholders can then be given the information to which they are entitled in the form that best suits their purposes and best encourages learning.

REFERENCES

Arnstein, S. 1969.. The ladder of citizen participation. *Journal of American Institute of Planners, 35,* 221.

Choudhary, A., & Tandon, R. n.d... *Participatory evaluation: Issues and concerns.* New Delhi, India: Society for Participatory Research in Asia.

Cronbach, L. J., & Associates. 1981.. *Toward reform of program evaluation: Aims, methods, and institutional arrangements.* San Francisco: Jossey-Bass.

Deshler, D. 1990.. Concept mapping. In J. Mezirow & Associates Eds.., *Fostering critical reflection in adulthood: A guide to transformative learning.* San Francisco: Jossey-Bass.

FAO 1990.. *Participation in practice: Lessons learned from the FAO People's Participation Programme.* Rome: Food and Agriculture Organization, People's Participation Programme.

Goldstein, I. L. 1986.. Evaluation procedures. In I. L. Goldstein, *Training in organizations: Needs assessment, development and evaluation* 2nd ed... Monterey, CA: Brooks/Cole.

Murphy, J., & Marchant, T. 1988.. *Monitoring and evaluation in extension agencies.* Washington, DC: The International Bank for Reconstruction and Development/The World Bank.

Bibliography

Ahmed, A. (1982). *The role of the information system in development.* Studies Series, No. 314. Baghdad, Iraq: Ministry of Culture and Information.

Arasingham, S. R. (1981). Agricultural extension in Sri Lanka. *Tropical Agriculturist, 137,* 41-54.

Black, A. W. (1976). *Organizational genesis and development: A study of Australian agricultural colleges.* St. Lucia: University of Queensland Press.

Blackburn, D. J., & Vist, D. L. (1984). Historical roots and philosophy of extension. In D. J. Blackburn (Ed.), *Extension handbook* (1st ed.). Guelph, Ontario: University of Guelph.

Bne Saad, M. H. A. Al-h. (1990). *An analysis of the needs and problems of Iraqi farm women: Implications for agricultural extension services.* Unpublished doctoral thesis, University College, Dublin.

Boulet, M. (n.d.). Un type original d'enseignant: Le professeur departmental d'agriculture. Unpublished paper.

Bray, P. (1984). *Joseph Needham's science and civilization in China: Vol. 6. Biology and biological technology. Pt. II: Agriculture.* Cambridge: Cambridge University Press.

Chambers, R. (1993). *Challenging the professions: Frontiers for rural development.* London: IT Publications.

CLVS (1845). Wiesenbewasserung. *Wochenblatt für Land-und Hauswirthschaft, Gewerbe und Handel* (Centralstelle des Landwirthschaftlichen Vereins zu Stuttgart), *24,* 133.

de Failly, D. (1970). Histoire de l'enseignement agricole au Congo (Pt. 1). *Cahiers Congolais, 13 (1),* 100-133.

Delman, J. (1991). *Agricultural extension in Renshou County, China: A case-study of bureaucratic intervention for agricultural innovation and change.* Aarhus, Denmark: Institute of East Asian Studies.

Elvin, M. (1973). *The pattern of the Chinese past.* London: Methuen.

FAO (1993). *The potentials of microcomputers in support of agricultural extension, education and training.* Rome: FAO.

Garforth, C. (1986). Mass media and communications technology. In G. E. Jones (Ed.), *Investing in rural extension: Strategies and goals* (p. 185-192). London and New York: Elsevier Applied Science Publishers.

Garforth, C. (1993). Seeing the people for the trees: Training for social forestry in Karnataka, India. *Rural Extension Bulletin, 2,* 33-39.

Gray, E. M. (1952). *The educational work of Philipp Emanuel von Fellenberg (1771-1844).* Unpublished master's thesis, The Queen's University, Belfast.

Grobel, M. (1933). *The Society for the Diffusion of Useful Knowledge, 1826-1846, and its relation to adult education in the first half of the nineteenth century.* Unpublished master's thesis. University of London.

Guggisberg, K. (1953). *Philipp Emanuel von Fellenberg und sein Erziehungsstaat* (2 vols.). Bern, Switzerland: Verlag Herbert Lang.

Hayward, J. (1990). Agricultural extension: The World Bank's experience and approaches. In FAO *Report of the Global Consultation on Agricultural Extension* (p. 115-134). Rome: FAO.

Hudson, K. (1972). *Patriotism with profit: British agricultural societies in the eighteenth and nineteenth centuries.* London: Hugh Evelyn.

Jenkins, H. M. (1884). *Report on agricultural education in north Germany, France, Denmark, Belgium, Holland, and the United Kingdom* (C.3981-1). London: Eyre and Spottiswoode.

Leeuwis, C. (1993). *Of computers, myths and modelling: The social construction of diversity, knowledge, information, and communication technologies in Dutch horticulture and agricultural extension.* Wageningen Studies in Sociology, 36. Wageningen: Agricultural University.

Logan, G. N. (1984). Man and the land in Queensland education 1874-1905: Agricultural education as a solution to man-physiography dissonance in colonist Queensland. Unpublished M.Ed. thesis, University of Queensland.

Lucas, C. P. (1913). *A historical geography of the British colonies: Vol. Ill, West Africa* (3rd ed.). Oxford: Clarendon Press.

Maier-Bode, F. (1910). *Die Organization und die Erfolge des landwirtschaftlichen Wanderunterrichts im Königreich Bayern.* Landsberg am Lech: Georg Verza.

Masefield, G. B. (1950). *A short history of agriculture in the British colonies.* Oxford: Clarendon Press.

Ministère de l'Agriculture (1882). *Bulletin,* 1st year, vol. 1.

Ministry of Agriculture (1993). *Outline of cooperative agricultural extension service in Japan.* Tokyo: Ministry of Agriculture, Forestry and Fisheries (Extension and Education Division).

Mook, B. T. (1982). *The world of the Indian field administrator.* New Delhi: Vikas Publishing House.

Moris, J. (1991). *Extension alternatives in tropical Africa.* London: Overseas Development Institute.

Perkins, D. H. (1969). *Agricultural development in China, 1368-1968.* Chicago: Aldine Publishing; Edinburgh: Edinburgh University Press.

Plunkett, H. (1901-02). Agricultural education for Ireland. *Department of Agriculture and Technical Instruction for Ireland, Journal,* 2(1), 18-41

Province of Ontario (1900). *Report of the superintendent of farmers' institutes of the Province of Ontario, 1899-1900.* Toronto: L. K. Cameron.

Report of the Recess Committee on the establishment of a Department of Agriculture and Industries for Ireland (1896). Dublin: Browne and Nolan.

Rivera, W. M., & Gustafson, D. J. (Eds.). (1991). *Agricultural extension: Worldwide institutional evolution and forces for change.* Amsterdam and New York: Elsevier.

Russell, E. J. (1966). *A history of agricultural science in Great Britain, 1620-1954.* London: George Alien and Unwin.

Smith, H. (1972). *The Society for the Diffusion of Useful Knowledge, 1826-1846.* London: Vine Press.

Smith, V. (1994, Summer). A learner centred approach to social skills for technical foresters. In *From the field.* Rural Development Forestry Network Paper 17e (p. 12-20). London: Overseas Development Institute.

Society for the Diffusion of Useful Knowledge (1827). *Rules of the Society/or the Diffusion of Useful Knowledge.* London: William Clowes.

Tajima, S. (1991). Processes of development of agricultural extension in Japan. In *Agricultural extension in Asia and the Pacific* (p. 19-36). Tokyo: Agricultural Productivity Organization.

True, A. C. (1895). Education and research in agriculture in the United States. In *Yearbook of the U.S. Department of Agriculture, 1894* (p. 81-116). Washington, DC: U.S. Government Printing Office.

Tusser, T. ([1580] 1984). *Five hundred points of good husbandry.* Oxford: Oxford University Press.

White, K. D. (1970). *Roman farming.* London: Thames and Hudson.

Willis, J. C. (1922). *Agriculture in the tropics: An elementary treatise* (3rd ed.). Cambridge: Cambridge University Press.

Zijp, W. (1994). *Improving the transfer and use of agricultural information: A guide to information technology.* World Bank Discussion Paper 247. Washington, DC: The World Bank.

Index